Gene Transcription

Mechanisms and Control

Robert J. White

Institute of Biomedical and Life Sciences
Division of Biochemistry and Molecular Biology
University of Glasgow
Glasgow, UK

Blackwell
Science

21203237

© 2001 by
Blackwell Science Ltd
Editorial Offices:
Osney Mead, Oxford OX2 0EL
25 John Street, London WC1N 2BS
23 Ainslie Place, Edinburgh EH3 6AJ
350 Main Street, Malden
   MA 02148-5018, USA
54 University Street, Carlton
   Victoria 3053, Australia
10, rue Casimir Delavigne
   75006 Paris, France

Other Editorial Offices:
Blackwell Wissenschafts-Verlag
GmbH
Kurfürstendamm 57
10707 Berlin, Germany

Blackwell Science KK
MG Kodenmacho Building
7–10 Kodenmacho Nihombashi
Chuo-ku, Tokyo 104, Japan

Iowa State University Press
A Blackwell Science Company
2121 S. State Avenue
Ames, Iowa 50014–8300, USA

First published 2001
Set by Graphicraft Limited,
Hong Kong
Printed and bound at the Alden
Press, Oxford and Northampton

A catalogue record for this title
is available from the British Library

ISBN 0-632-04888-3

Library of Congress
Cataloging-in-publication Data
White Robert, J., 1963–
   Gene transcription : mechanisms
and control / Robert J. White.
      p. cm.
   Includes index.
   ISBN 0–632–04888–3
   1. Genetic transcription.   2.
Genetic transcription—Regulation.
I. Title.
QH450.2.W48 2000
572.8′845—dc21   00–040330

DISTRIBUTORS

Marston Book Services Ltd
PO Box 269
Abingdon, Oxon OX14 4YN
(Orders: Tel: 01235 465500
         Fax: 01235 465555)

USA
Blackwell Science, Inc.
Commerce Place
350 Main Street
Malden, MA 02148-5018
(Orders: Tel: 800 759 6102
              781 388 8250
         Fax: 781 388 8255)

Canada
Login Brothers Book Company
324 Saulteaux Crescent
Winnipeg, Manitoba R3J 3T2
(Orders: Tel: 204 837 2987)

Australia
Blackwell Science Pty Ltd
54 University Street
Carlton, Victoria 3053
(Orders: Tel: 3 9347 0300
         Fax: 3 9347 5001)

For further information on
Blackwell Science, visit our website:
www.blackwell-science.com

Contents

Preface

Transcription is an essential process which is required for abstract genetic information to become physical reality. As such it is fundamental to every living organism. It impinges upon all biological processes, such as growth, development and the ability to respond to environmental circumstances. Defects in transcription are known to characterise the majority of human diseases.

Given these truths, it is clear that any serious student of the life sciences must have some familiarity with the principles of transcription. Furthermore, that familiarity needs to be maintained and regularly updated, since the field moves rapidly, as befits its place at the cutting edge of biological disciplines. The aim of this book is to provide a non-expert with the established fundamentals of transcription, as well as some of the most exciting recent developments. It was completed in January 2000 and therefore reflects the field as it was understood by the author at that point in time. The intended audience includes researchers who work outside of transcription, but want to keep abreast of current thinking in the field, teachers and students of final year undergraduate courses in gene expression, and postgraduate students who require an introduction to transcription as a starting point for future studies. A very basic introduction has been provided in Chapter 1; this will prove too elementary for the majority of readers, who may prefer to skip to Chapter 2, which continues at a more sophisticated level that is maintained throughout the remainder of the book. Scattered through the chapters and boxed away from the main text, I have described some of the key experimental approaches that have been often used to investigate this topic.

Like any highly technical science, the field of transcription is rife with jargon. I have attempted to avoid this to the best of my abilities. However, it is impossible to describe a subject such as transcription without regularly employing many specialized terms. I hope that readers can live with this; in the absence of such terms, descriptions rapidly become vague, banal and bereft of detail. Nevertheless, I have purposefully avoided mentioning many of the transcription factors, preferring instead to illustrate the principles using a select group of representative examples, which reappear regularly in different contexts throughout the pages of this work. Some readers may be offended to find that their favourite factor has been neglected. I can only apologise, but the book was not intended to provide an exhaustive catalogue. I have also focussed almost entirely on animals and fungi, largely ignoring the interesting worlds of plants, prokaryotes and archaea. This stategy was adopted because of the spatial constraints of a book of this size; I felt it would be more rewarding to cover selected systems in adequate detail, rather than a superficial reference to all life forms. Inevitably, the choice of examples is somewhat subjective, reflecting my own personal interests and expertise. To illustrate the importance of transcription in development, I have chosen to focus on *Drosophila* embryogenesis, as this is probably the best characterized system. Elsewhere, I have referred very frequently to human disease states that can arise through defects in transcription. Examples include cardiovascular disorders, diabetes and many types of cancer. Indeed, the book regularly returns to the theme of cancer, which invariably involves aberrations of transcription. For some years now I have been absorbed by transcription and its many connections with human diseases. I hope very much that I shall be able to pass on to my readers some of this enthusiasm.

Robert J. White

Abbreviations

A	adenine
ACF	ATP-utilizing chromatin assembly and remodelling factor
ACTR	activator of the TR and RAR
AD	activation domain
AF	activation function
AIB	amplified in breast cancer
APC	adenomatous polyposis coli
bHLH	basic helix-loop-helix
bp	base pair
BrdU	bromodeoxyuridine
BRF	TFIIB-related factor
C	cytosine
CAK	cdk activating kinase
cAMP	cyclic AMP
CAT	chloramphenicol acetyltransferase
CBP	CREB-binding protein
cdk	cyclin-dependent kinase
C/EBP	CCAAT/enhancer binding protein
CF	cleavage factor
CF	core factor
ChIP	chromatin immunoprecipitation
CHO	Chinese hamster ovary
CHRAC	chromatin remodelling and assembly complex
Ci	cubitus interruptus
CKII	casein kinase II
CPSF	cleavage-polyadenylation specificity factor
CRE	cAMP-responsive element
CREB	cAMP-response element-binding protein
CS	Cockayne's syndrome
CstF	cleavage stimulation factor
CTD	C-terminal domain
DBD	DNA-binding domain
DSE	distal sequence element
dTCF	*Drosophila* TCF
EGF	epidermal growth factor
eIF2	eukaryotic translation initiation factor 2
ER	oestrogen receptor
FACT	facilitates chromatin transcription

G	guanine
GR	glucocorticoid receptor
GSK3	glycogen synthase kinase 3
GST	glutathione *S*-transferase
GTP	guanosine triphosphate
HAT	histone acetyltransferase
HCF	host-cell factor
HDAC	histone deacetylase
HHV8	human herpesvirus 8
HIV	human immunodeficiency virus
HLH	helix-loop-helix
HMG	high mobility group
HNF	hepatocyte nuclear factor
HPV	human papillomavirus
HSF	heat-shock factor
HSV	herpes simplex virus
HTH	helix-turn-helix
HTLV-1	human T-cell leukaemia virus type 1
ICAM-1	intracellular adhesion molecule 1
ICER	inducible cAMP early repressor
Id	inhibitor of differentiation
IκB	inhibitor of NF-κB
IKK	IκB kinase
IL	interleukin
ISWI	imitation switch
JAK	Janus kinase
JNK	Jun N-terminal kinase
KID	kinase-inducible domain
KSHV	Kaposi's sarcoma-associated herpesvirus
LBD	ligand-binding domain
LCR	locus control region
LEF	lymphoid enhancer factor
MAPK	mitogen-activated protein kinase
MBD	methyl-CpG-binding domain
MBT	midblastula transition
MCK	muscle creatine kinase
MEF2	myocyte-enhancer binding factor
ML	major late
MMTV	mouse mammary tumour virus
MNase	micrococcal nuclease
MOZ	monocytic-leukaemia factor
mRNA	messenger RNA
NCoR	nuclear receptor corepressor
N_3RdU	[*N*-(*p*-azidobenzoyl)-3-aminoallyl]-deoxyuridine

NER	nucleotide excision repair
NLS	nuclear localization signal
NMR	nuclear magnetic resonance
NURD	nucleosome remodelling histone deacetylase complex
NURF	nucleosome remodelling factor
PABII	poly(A)-binding protein II
PAP	poly(A) polymerase
P/CAF	P300/CBP-associated factor
PCR	polymerase chain reaction
PDGF	platelet-derived growth factor
PKA	protein kinase A
PLZF	promyelocytic leukaemia zinc finger
PML	promyelocytic leukaemia
pols	RNA polymerases
POU_H	POU homeodomain
POU_s	POU-specific domain
PPARγ	peroxisome proliferator-activated receptor gamma
PR	progesterone receptor
PSE	proximal sequence element
PTF	PSE-binding transcription factor
RA	retinoic acid
RAP	RNA polymerase associated
RAR	retinoic acid receptor
rDNA	ribosomal DNA
RNA	ribonucleic acid
rRNA	ribosomal ribonucleic acid
RNP	ribonucleoprotein
RSC	remodelling the structure of chromatin
RXR	retinoid X receptor
SAGA	Spt-Ada-Gcn5-acetyltransferase
SAPK	stress-activated protein kinase
scs	specialized chromatin structures
SH2	src homology 2
SINE	short interspersed repeat
SMRT	silencing mediator for RAR and TR
SNAPc	snRNA activating protein complex
snRNA	small nuclear RNA
SRB	suppressor of RNA polymerase B
SRC-1	steroid receptor coactivator-1
SRE	serum response element
SRF	serum response factor
STATs	signal transducers and activators of transcription
SV40	simian virus 40
T	thymine

TAF	TBP-associated factor
TAK	Tat-associated kinase
TBP	TATA-binding protein
TCF	T-cell factor
TCF	ternary complex factor
TdT	terminal deoxynucleotidyl transferase
T_H cell	T helper cell
TNF	tumour necrosis factor
tRNA	transfer RNA
TTD	trichothiodystrophy
U	uracil
UAF	upstream activation factor
UAS_G	galactose upstream activating sequence
UBF	upstream binding factor
UCE	upstream control element
uORFs	upstream open reading frame
VHL	von Hippel–Lindau
XP	xeroderma pigmentosum

Chapter 1: Introduction

Gene expression

The nature of a living organism is determined by its genes. Humans have different genes from mice, which is why people are different from mice. Most of the cells within a person have the same set of genes. Yet the cells within a person can be quite different from one another; for example, skin cells are flat and translucent whereas erythrocytes are round and red. How can a person's cells vary when they have the same genes? The answer is that what influences a cell's identity is not just which genes it possesses, but which of these are expressed. A cell will only express a fraction of its genes at any given time. Furthermore, the genes which are expressed vary from one cell type to another. Thus, a liver cell is different from a skin cell because it expresses a different combination of genes out of the total set that they have in common. Only those genes that are expressed contribute directly to a cell's characteristics or phenotype. A useful analogy is to think of a gene as a book, in which case expressing that gene is akin to reading the book, thereby unlocking its latent content.

Genes that encode essential structural or metabolic components of the cell may be expressed continuously in most or all cell types; these are referred to as 'housekeeping' genes. By contrast, others are much more specialized and may only be expressed under particular circumstances or in a single tissue. Gene expression can vary during the lifetime of a cell and this allows it to respond to changes in its environment. For example, when skin cells are exposed to strong sunlight, they may respond by expressing the genes that are responsible for pigmentation. Development is accompanied by dramatic changes in gene expression, such that a stem cell may express a very different set of genes from its differentiated progeny. This allows organisms to generate a range of different cell types from a single set of genes.

In molecular terms, what does it mean to 'express' a gene? Genetic information is stored as DNA. DNA is a very stable storage molecule, which is important because genetic information must be preserved from one generation to the next. However, it is also a rather passive molecule. For the information stored within the DNA to be utilized then, it must be transferred into a molecular form with a more active role. This is what is meant by 'expressing' a gene. The first step in gene expression is to produce RNA molecules by the process of transcription. RNA can perform a variety of structural and catalytic processes within the cell. The precise nature of an RNA molecule is determined by the gene that gives rise to it. Although RNA can carry out many functions, it lacks the chemical complexity that is required for most of the jobs that keep a cell alive. Instead, RNA often serves as a template for the synthesis of protein, a much more versatile type of molecule. Proteins carry out most of the catalytic and structural roles within an organism. The identities of the proteins produced are determined by the types of RNA that serve as templates for their synthesis, in a process called translation. The choice of proteins synthesized is therefore set by which RNAs are present to serve as translational templates, and this is dictated by which genes are transcribed. In other words, the nature of a cell is determined by which proteins and RNAs are produced, which is in turn decided by the genes that are expressed. Thus, gene expression is the fundamental determinant of an organism's nature and identity.

Because it is the first step in gene expression, transcription is subject to a wide variety of regulatory controls. Although gene expression can also be regulated post-transcriptionally, the most common and important level of control is transcriptional. The regulation of transcription is intimately involved in almost every biological process. It underlies growth and development. It allows the cell to constantly adapt in response to environmental changes and metabolic requirements. It serves as a dynamic link between the abstract informational content of an organism's genome and its concrete physical properties. Defects in transcription are responsible for a wide range of human diseases, as befits its central importance in sustaining life. For example, at least 300 different genes were found to be aberrantly transcribed in primary colon cancers. Indeed, transcriptional defects play a crucial part in the development of all kinds of cancer. They are also associated with many other types of disease, including cardiovascular hypertension, diabetes and many developmental abnormalities. Many of the links with human disease will be described during the course of this book. But first, the remainder of this chapter will provide a brief and elementary background. Many readers may prefer to skip this introduction and go straight to Chapter 2.

DNA

Genes are made out of a polymeric substance called **d**eoxyribo**n**uleic **a**cid (DNA). This is composed of three relatively simple types of compound, viz, heterocyclic nitrogenous bases, the sugar deoxyribose, and phosphoric acid (Fig. 1.1). Combinations of four different bases are used; two of these, cytosine (C) and thymine (T), have a six-membered ring and are classified as pyrimidines, whereas the other two, adenine (A) and guanine (G), have a five-membered ring fused to a six-membered ring and are designated as purines. Despite their simplicity, these bases provide the alphabet in which the genetic information is written. In DNA they are coupled to a sugar phosphate backbone via N-glycosidic linkages with the deoxyribose moities (Fig. 1.2). These chemical linkages can be repeated millions of times to create enormous polymeric chains. Each of these chains is found associated with a complementary

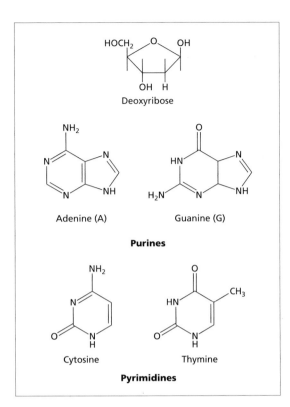

Fig. 1.1 Chemical structures of deoxyribose and the purine and pyrimidine heterocyclic bases that, along with phosphoric acid, provide the building blocks of DNA.

chain of the same size that is orientated in the opposite direction. The two chains are referred to as strands and they are held together by hydrogen bonds between the two sets of bases. Adenines on one strand form two hydrogen bonds with thymines on the opposite strand and cytosines on one strand form three hydrogen bonds with guanines on the other strand (Fig. 1.3). The base sequence of a strand of DNA can therefore be predicted from the sequence of its complementary strand. At physiological ionic strength and temperatures, the two strands of a DNA molecule adopt a regular double helical configuration (Fig. 1.4). A complete turn of the helix is made every 3.4 nm and requires 10.5 base pairs (bp). This means that each base is rotated ~34° around the central axis of the double helix, relative to its immediate neighbours. Two grooves in the molecule allow proteins to gain access to the bases. The major groove has a width of 2.2 nm whereas the minor groove is only 1.2 nm wide. The DNA molecule has a diameter of 2 nm but may be more than a metre in length.

Transcription

As already explained, in order to express its genetic information, DNA must be transcribed to generate molecules of **ribo**nucleic **a**cid (RNA). RNA is similar chemically to DNA, except that it contains ribose instead of deoxyribose and

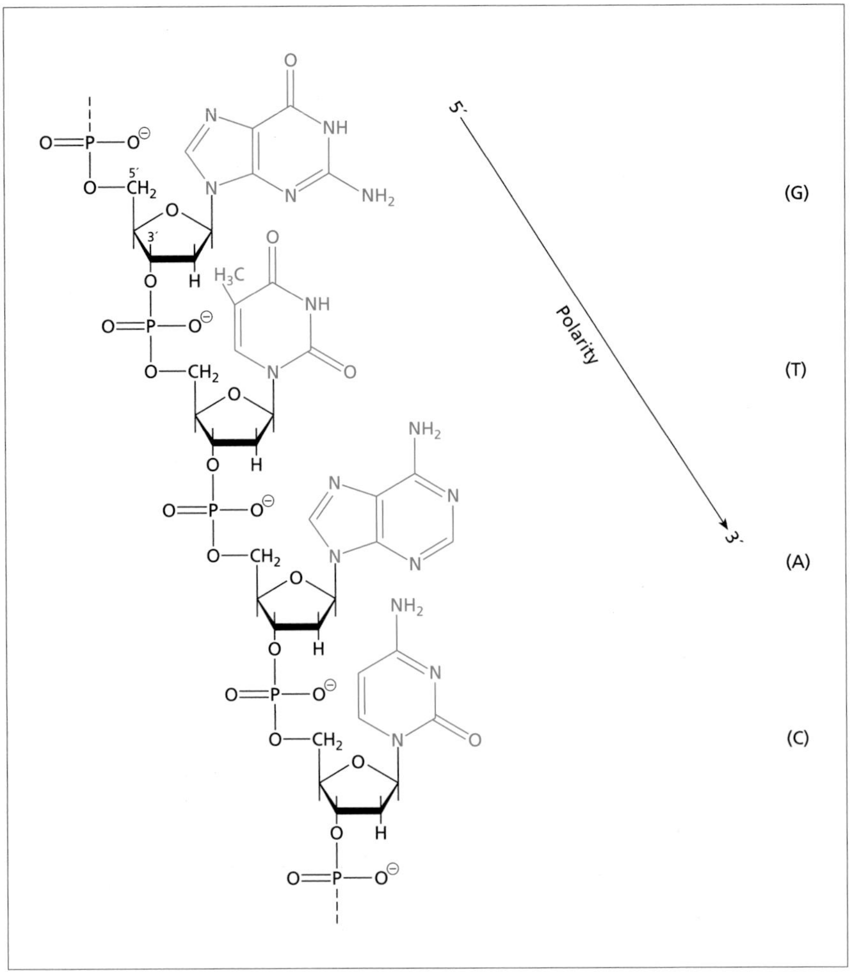

Fig. 1.2 Covalent linkage of heterocyclic bases with an alternating sugar-phosphate backbone to produce a tetranucleotide fragment of DNA with the sequence GTAC.

uracil (U) instead of thymine (Fig. 1.5). Transcription is the process of RNA synthesis and is carried out by large enzymes called RNA polymerases. They use one strand of a DNA molecule as a template to generate RNA copies in which the order of bases is complementary to the template strand and identical to the opposite DNA strand (Fig. 1.6). RNA polymerases begin transcription at the initiation or start site of a gene, which is often denoted as nucleotide +1. Sequences prior to the start site are described as being upstream of it, whereas those after the start site are referred to as downstream. Having initiated transcription at +1, an RNA polymerase moves downstream along the gene, synthesizing an RNA transcript as it goes. Eventually it reaches a termination site, at which point it releases the transcript and dissociates from the DNA.

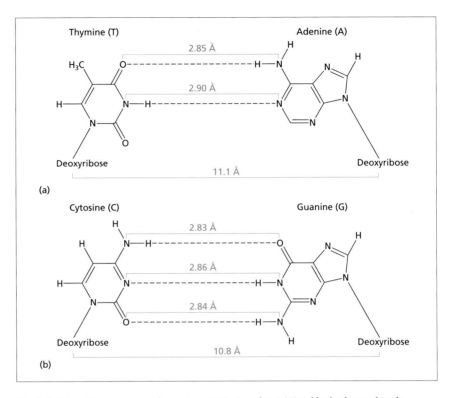

Fig. 1.3 Pairs of bases on complementary DNA strands are joined by hydrogen bonds. Adenines always pair with thymines and guanines always pair with cytosines. Two hydrogen bonds are formed by the A/T pair whereas three are formed by the G/C pair. The lengths of these bonds are indicated. Because the A/T pair is only joined by two hydrogen bonds, it is easier to separate and therefore less stable than a G/C pair. The hydrogen bonds between base pairs determine the specificity of transcription and DNA replication.

It is important to note that genes are only part of a DNA molecule; they may be separated by extensive regions of DNA that are not transcribed, especially in the large genomes of higher organisms. Even though the non-transcribed regions do not code for RNA, they often contain important regulatory signals that help direct RNA polymerases to the start sites of genes. The region of DNA immediately upstream of a gene usually contains especially important regulatory signals and is referred to as a promoter. More remote regulatory regions of DNA are often called enhancers. Promoters and enhancers will be described in Chapters 4–7.

In order to transcribe, RNA polymerases must separate the two strands of DNA by breaking the hydrogen bonds between its bases. A/T pairs, which are joined by two hydrogen bonds, are easier to separate than G/C pairs, which have three. The region at which the DNA strands are separated is referred to as a transcription bubble and is the site of RNA synthesis. Within this bubble, the nascent RNA forms base pairs with the template DNA strand. The fidelity of transcription is determined by the RNA polymerase, which ensures that each

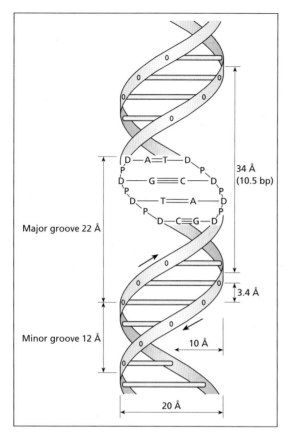

Fig. 1.4 The double helix formed by a DNA molecule. The two strands run in opposite directions. The phosphodiester backbones (P, phosphate; D, deoxyribose) are on the outside of the double helix whereas the bases (A, G, T, C) face inwards. Hydrogen bonds form between complementary bases on the opposite strands. Proteins and chemicals can gain access to the bases through the major and minor grooves. The double helix has a diameter of 2 nm. It makes a complete turn every 3.4 nm and this requires 10.5 bp. These dimensions are characteristic of the most common DNA conformation under physiological conditions, which is referred to as B form. Other conformations with slightly different characteristics can also occur.

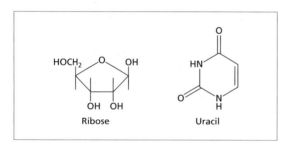

Fig. 1.5 Chemical structures of ribose and uracil, which are found in RNA in place of deoxyribose and thymine, respectively, which are present in DNA.

new ribonucleotide must pair with a base in the template, according to the rule that G pairs with C, C pairs with G, A pairs with U and T pairs with A. Ribonucleotides that are not complementary to the template base are expelled from the active site, allowing another candidate to enter. Once a match is found, the new ribonucleotide is attached covalently to the growing RNA chain, in a condensation reaction that is catalysed by the RNA polymerase. The 3'-OH group of the last ribonucleotide in the growing RNA chain attacks the triphosphate group of the incoming ribonucleotide; the terminal two phosphate groups (β and γ) are released as pyrophosphate, whereas the α

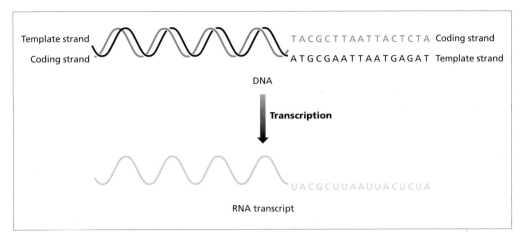

Fig. 1.6 Transcription makes use of one DNA strand as template to generate an RNA transcript that is complementary in sequence to the template DNA strand and identical in sequence to the coding DNA strand (except that U in RNA replaces T in DNA).

group forms part of the new phosphodiester bond that incorporates the base into the growing chain (Fig. 1.7). This reaction occurs in the catalytic site of the RNA polymerase.

As the polymerase progresses along the gene, the transcription bubble moves with it and the RNA chain grows longer (Fig. 1.8). However, the DNA duplex reforms behind it, displacing the transcript so that the bubble remains a fairly constant size of ~12–20 bp. Only two or three bases of RNA at its growing tip are actually hybridized to the template DNA, so much of the unwound DNA within the transcription bubble must be in a single-stranded state. Once it is displaced from its transient interaction with the DNA, the newly formed RNA is thought to be held temporarily in a high-affinity RNA-binding site within the polymerase. This site retains about 25 ribonucleotides at the growing end of the nascent RNA chain. The RNA polymerase continues in this fashion until the end of the gene is reached, at which point it releases the completed primary transcript.

RNA processing

Many different types of RNA are found within a cell. The principle categories are the **m**essenger **RNA** (mRNA), **r**ibosomal **RNA** (rRNA) and **t**ransfer **RNA** (tRNA). An mRNA provides the template for protein synthesis, dictating the sequence of the polypeptide chain which it encodes. The process of protein synthesis is referred to as translation and is carried out by ribosomes, which are large macromolecular assemblies. In addition to many proteins, key components of ribosomes are the rRNAs, which perform both structural and catalytic roles in translation. The building blocks of proteins are amino acids,

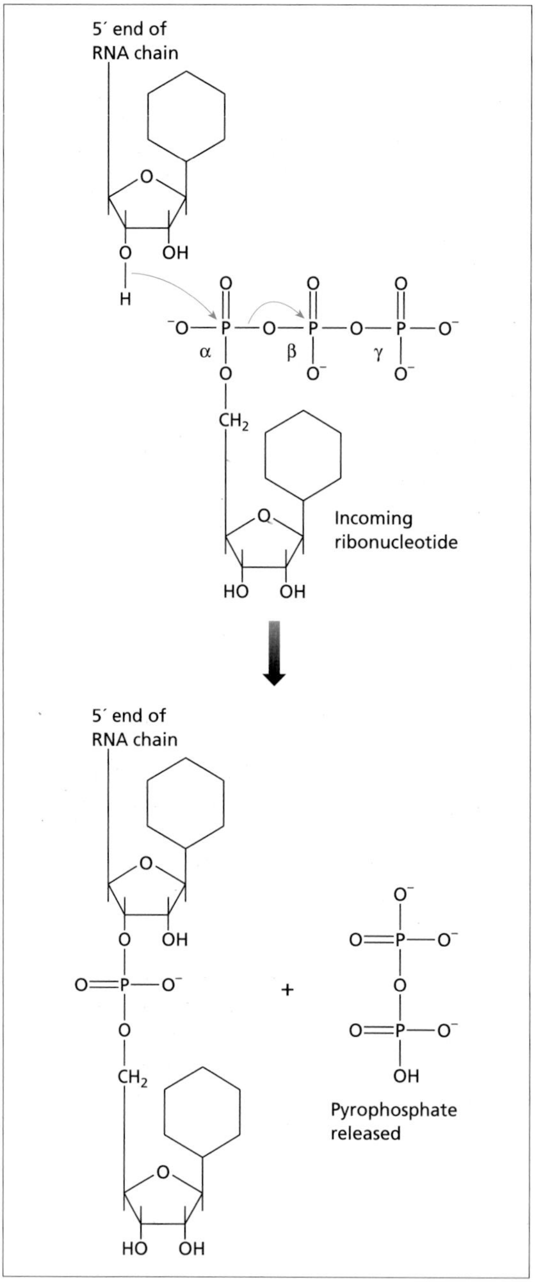

Fig. 1.7 RNA synthesis is achieved by a chemical reaction in which the 5′ triphosphate of the incoming ribonucleotide undergoes hydrophilic attack by the 3′-OH group of the most recently added ribonucleotide in the growing chain. A new phosphodiester bond is formed and pyrophosphate is released.

which are recruited into nascent polypeptides with the help of tRNAs. Translation will be described in a little more detail below.

To generate mature mRNA, rRNA or tRNA molecules, primary transcripts must undergo a series of covalent modifications. The nature of these post-transcriptional processing reactions depends very much on the type of RNA.

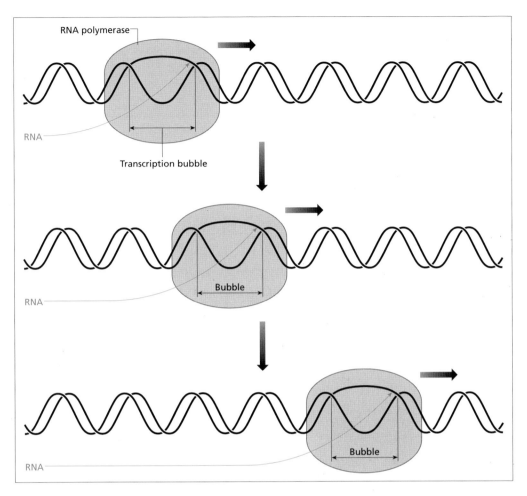

Fig. 1.8 RNA synthesis takes place in a 'transcription bubble' of unwound DNA, in which 12–20 bp of the duplex are dissociated. RNA polymerase catalyses the unwinding of the DNA duplex and the addition of new bases to the elongating transcript. The identity of each new base is dictated by pairing with the corresponding base in the unwound region of the template DNA strand. As the RNA polymerase progresses downstream, the transcription bubble moves with it and the DNA duplex reforms behind it, displacing the nascent transcript from the DNA–RNA hybrid.

For example, pre-mRNA molecules have a **g**uanosine **trip**hosphate (GTP) nucleotide added to their 5′ ends to form a structure called a cap that facilitates translation. In addition, stretches of ~200 adenine residues are added to their 3′ ends in a reaction called polyadenylation, which stabilizes the transcript. The other major processing step for pre-mRNA molecules is splicing. In many eukaryotic genes, the protein-coding regions (exons) are interrupted by non-coding regions called introns. Only a minority of genes are organized in this way in lower eukaryotes such as *Saccharomyces*, but the vast majority of protein-coding genes are discontinuous in higher organisms. A typical example

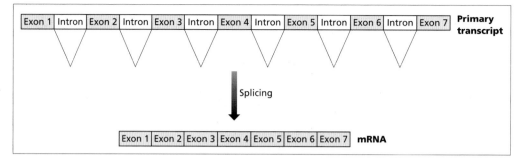

Fig. 1.9 The process of splicing. Many pol II-transcribed genes contain non-coding regions called introns as well as the exons which code for protein sequence. Primary transcripts are processed to remove the intron-derived regions from the RNA in a series of splicing reactions. After each intron copy is excised, the exonic sequences are joined together so that the protein-coding region is no longer interrupted in the processed mRNA.

in mammals will have 7–8 exons spread out over ~16 kb of DNA; the exons will be relatively short (~100–200 bp) compared to the introns (~1 kb). The primary transcript is produced by copying the entire gene, but the regions encoded by introns must be processed out in order to generate a mature mRNA. The intronic sequences are excised from the RNA and the exonic sequences are joined together in the same order as they are found in the gene; this reaction is referred to as splicing (Fig. 1.9). Chapters 4 and 13 will describe in much more detail the post-transcriptional processing of pre-mRNA molecules. Some tRNA molecues also undergo splicing; in addition, they are processed at their 5′ ends and many of their bases undergo covalent modifications. The bases of rRNA are also heavily modified prior to incorporation into ribosomes. Furthermore, most of the rRNA is synthesized as a single large precursor molecule from which three separate rRNAs are excised by a series of endonucleolytic cleavages.

Translation

Before it can perform its function, mRNA must be transported out of the nucleus, where it is synthesized and processed, and into the cytoplasm. There it can be bound by ribosomes and serve as template for polypeptide synthesis (Fig. 1.10). The nature of the protein encoded by an mRNA is dictated by its base sequence, which was of course determined by the DNA template on which it was synthesized. In essence, the bases of DNA and RNA can be regarded as a four-letter alphabet in which the genetic information is written. The protein synthetic apparatus reads this information by interpreting the base sequence so as to specify a particular order of amino acids in a polypeptide chain. The alphabet is read in triplets of bases, each of which is called a codon, but might be regarded as a word if we pursue the analogy. Every possible triplet specifies

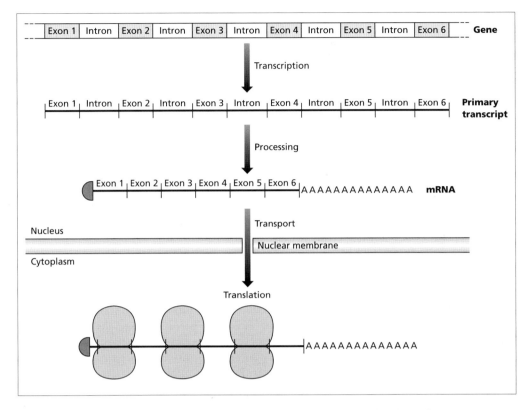

Fig. 1.10 Steps between DNA and protein. DNA is transcribed in the nucleus to generate a primary RNA transcript that must undergo a series of processing reactions. For mRNA, these involve capping at the 5′ end, cleavage and polyadenylation at the 3′ end and the splicing out of the sequence encoded by introns. The mature mRNA is then exported to the cytoplasm where it associates with ribosomes and is translated into protein.

a particular amino acid (Table 1.1). A ribosome travels along an mRNA and adds a new amino acid to the nascent polypeptide as it passes each triplet. Codon recognition is achieved by tRNA molecules, which associate with the ribosome and form hydrogen bonds with the trinucleotides in the mRNA. Every codon can be matched to a base triplet that is referred to as an anticodon, within a particular tRNA. The opposite end of the tRNA is coupled covalently through an ester bond to the carboxyl group of an amino acid (Fig. 1.11). Each amino acid is carried by a different tRNA with a distinct anticodon sequence. Some amino acids can be specified by several different codons and these are therefore carried by multiple tRNAs with anticodons that match each of the relevant codons. By base-pairing with the mRNA triplets, the tRNAs ensure that the order in which amino acids are added to the growing polypeptide is dictated by the sequence of the mRNA and therefore by a gene. This process, in which genetic information is decoded so as to direct the synthesis of proteins, is catalysed by ribosomes and is referred to as translation.

Table 1.1 The genetic code.

	U	C	A	G	
U	PHE	SER	TYR	CYS	U
	PHE	SER	TYR	CYS	C
	LEU	SER	STOP	STOP	A
	LEU	SER	STOP	TRP	G
C	LEU	PRO	HIS	ARG	U
	LEU	PRO	HIS	ARG	C
	LEU	PRO	GLN	ARG	A
	LEU	PRO	GLN	ARG	G
				ARG	
A	ILE	THR	ASN	SER	U
	ILE	THR	ASN	SER	C
	ILE	THR	LYS	ARG	A
	MET	THR	LYS	ARG	G
G	VAL	ALA	ASP	GLY	U
	VAL	ALA	ASP	GLY	C
	VAL	ALA	GLU	GLY	A
	VAL	ALA	GLU	GLY	G

Every triplet of mRNA bases (codon) can be interpreted as specifying a particular amino acid. The first letter of the codon is in the left hand column, the second is in the horizontal axis, and the third is in the right hand vertical column. For example, CAG codes for GLN (glutamine).

Transcription factors

Whereas prokaryotes employ a single RNA polymerase for all their transcription, eukaryotic nuclei contain three. These are named RNA **polymerases** (pols) I, II and III and each transcribes a different set of essential genes. Pol I synthesizes large rRNA, pol II synthesizes mRNA and most small nuclear **RNA** (snRNA, which is involved in splicing), and pol III synthesizes a variety of small stable RNAs, including tRNA and the smallest rRNA (5S rRNA). In actively growing cells, pol I can be responsible for ~70% of all nuclear transcription, whereas pols II and III account for ~20% and 10%, respectively. Despite its very high activity, pol I only transcribes a single type of gene, which encodes rRNA and is present in multiple copies. By contrast, by far the most diverse set of genes is transcribed by pol II. The templates transcribed by pols I, II and III are referred to as class I, class II or class III genes, respectively.

The pols are large enzymes containing many different subunits. As will be described in Chapter 2, several of the subunits are shared between two or three pols, whereas others are unique to an individual pol. The largest two subunits of each pol are unique, but share substantial homology between pols I, II and III, and also with the prokaryotic pol. These homologies provide an indication as to how the tripartite eukaryotic system evolved. Functions have

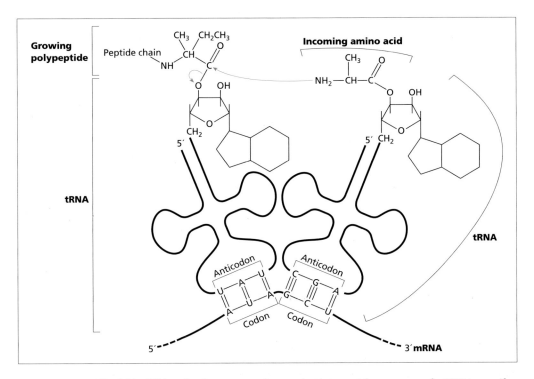

Fig. 1.11 tRNA molecules serve as adaptors that interpret the sequence of mRNA to specify a particular order of amino acid residues during polypeptide synthesis (translation) within a ribosome. As it is translated, each codon of an mRNA specifies a particular tRNA because of base-pairing with the latter's anticodon sequence. A specific amino acid will be covalently linked to the 3′ end of the tRNA and will therefore be brought into immediate proximity of the growing polypeptide. The nascent polypeptide is then transferred to the incoming amino acid, releasing the previous tRNA and increasing the length of the peptide chain by one residue. Synthesis of the new peptide bond is carried out by a peptidyl transferase activity within a ribosome. As translation proceeds, the mRNA passes through the ribosome, with each consecutive codon dictating the addition of a new amino acid.

been ascribed to several subunits, but the roles of many remain to be determined. Despite their great complexity, the pols on their own are unable to recognize genes and a purified pol will transcribe DNA randomly. Specificity is conferred on the process of transcription by proteins called transcription factors. It has been estimated that the mammalian genome may encode as many as 10 000 transcription factors. Many of these can recognize specific DNA sequences and thereby locate genes. They bind to target sequences, either directly or via other transcription factors, and thereby assemble complexes which serve as molecular beacons in order to recruit pols to the appropriate sites. Chapter 3 will describe the molecular details of DNA sequence recognition by a range of different transcription factors; this will involve a structural analysis of their DNA-binding domains and how these interact with specific bases in DNA.

Once it has been established how factors recognize appropriate gene sequences, Chapter 4 will explain how this can serve to recruit pol II to the starts of genes. It will go on to describe how transcripts initiate, elongate and terminate, giving rise to mRNA and snRNA molecules. These processes require a complex array of proteins, many of which have been highly conserved through evolution. Despite the elaborate polypeptide machinery that participates in basal pol II transcription, the process is highly inefficient. An additional class of transcription factors is therefore required to ensure that mRNA is synthesized at rates that are sufficient to sustain life. Indeed, the simple yeast *Saccharomyces cerevisiae* uses approximately 200 transcriptional activators to control its ~6200 pol II-transcribed genes. Chapter 5 will describe how such stimulatory factors operate to maintain adequate levels of expression. Chapters 6 and 7 will then describe transcription by pols I and III, respectively, both of which employ unique sets of transcription factors with their own interesting peculiarities. Overall, however, the machinery employed by pols I and III is less complex than that required by pol II. Because of this relative simplicity, many pioneering studies on the mechanics of transcription were carried out using pol I or pol III; the lessons learnt with these systems were often found subsequently to apply also to the more complicated process of pol II transcription.

The ~50 000–100 000 genes that belong to a human are contained within 6×10^9 base pairs of DNA, which has a total length of ~2 m. This must be packaged into a nucleus that is 6–8 μm in diameter. The packaging and organization is achieved primarily by histone proteins to produce a complex structure called chromatin. Chromatin is found in all eukaryotes and its organization plays a pivotal role in gene expression, in some cases facilitating and in other instances repressing. Chapter 8 will examine how chromatin influences the transcription of eukaryotic genes.

As explained at the beginning of this chapter, the characteristics of a cell are dictated by the genes which it expresses and this is controlled primarily at the level of transcription. Precise measurements in *S. cerevisiae* have revealed that expression levels of individual genes can vary over three orders of magnitude. The most highly expressed protein-coding genes may maintain 200 mRNA copies per yeast cell; examples include genes encoding ribosomal proteins or enzymes involved in energy metabolism, such as glyceraldehyde-3-phosphate dehydrogenase. By contrast, other mRNA molecules may average fewer than one copy per yeast. In a study of human cells, the most abundant mRNAs were found to be present in ~5300 copies per cell, whereas the rarest were present in one copy or fewer. To a substantial degree, this dramatic variation in expression is determined by the activity of transcription factors. Controlling the abundance of regulatory transcription factors provides a simple and economical method for dictating the expression of specific genes. For example, the albumin gene requires a factor called hepatocyte nuclear factor 1 (HNF-1) for high level expression; HNF-1 is much more abundant in liver than in most other cell types, which means that rapid albumin synthesis is restricted to the liver. This simple mechanism of control is probably the most common method for achieving tissue-specific gene expression. The production of transcription factors can be regulated in a variety of ways, as will be illustrated in Chapter 9. Furthermore, the presence of a transcription factor within a particular cell does not guarantee that it is functioning. Many factors are held in the cytoplasm, where they are unable to operate, until appropriate signals trigger their passage into the nucleus. Chapter 10 will provide several examples of this important phenomenon. Even when present in the nucleus, many factors require activation by post-translational modification or through interactions with other molecules. Examples of such regulatory mechanisms, which are described in Chapter 11, include phosphorylation, dimerization and the binding of ligands such as steroids.

Proliferating cells go through a repetitive cycle of replication and division. This sequence can have a profound effect on transcription. For example, some genes are only transcribed during particular periods of the cell cycle, such as many of the genes encoding the DNA replication apparatus. Furthermore, there is a general inhibition of transcription when higher eukaryotic cells undergo mitosis. Chapter 12 will examine the molecular basis of these phenomena. It will also describe how certain transcription factors can control the progress of the cell cycle, halting it when proliferation is disadvantageous. Two of these factors have been classified as tumour suppressors and their inactivation is a principal cause of many types of cancer.

Transcription can be strongly influenced by other nuclear events and can also have a profound effect upon some of these. Chapter 13 will describe the interactions between transcription, RNA processing, DNA replication and DNA repair. A clear illustration of the influence that transcription can have on DNA replication is provided by the fact that silent genes are replicated much later than genes that are actively expressed. A range of viral and cellular transcription factors have been shown to stimulate the initiation of DNA replication. Conversely, a replication fork has a decisive effect on expression and its passage will erase transcription complexes from genes. The post-transcriptional fate of an RNA is largely dictated by the polymerase responsible for its synthesis. Thus, pol II transcripts generally undergo capping, splicing and polyadenylation, whereas most of the products of pols I and III do not. Recent evidence has shown that much of the processing apparatus interacts physically with the pol II enzyme and is thereby targetted to premRNA products. The pol II-transcribed genes are also targetted for preferential repair following certain types of DNA damage. This may provide a substantial benefit to the organism, as damage to single-copy essential genes could be catastrophic. The targeting of transcribed DNA for rapid repair appears to reflect the fact that one of the key transcription factors that is used by pol II is also a component of the DNA repair apparatus. Mutations in this factor can result in several human genetic diseases, such as xeroderma pigmentosum which involves growth retardation, extreme sensitivity to ultraviolet light and a strong predisposition to skin cancer; these symptoms are likely to reflect defects in both transcription and DNA repair.

The switching on and off of genes is a central determinant of how organisms develop. Inevitably, this is dictated by changes in the levels and activities of key transcription factors. The final chapter will examine in detail a good example of this process, focusing on the fruitfly *Drosophila* as a well-characterized model. Many of the principles introduced in earlier chapters will be illustrated in this exciting biological context, where a hierarchical cascade of regulatory transcription factors is responsible for driving and coordinating embryonic development.

Chapter 2: The Nuclear RNA Polymerases

The transcription of DNA is carried out by RNA polymerases. Whereas a single enzyme is responsible for this job in prokaryotes and archaebacteria, eukaryotes have three RNA polymerases to share the task of transcribing the nuclear genes. An additional RNA polymerase is found in mitochondria, which carry a small DNA molecule of their own. This division of labour may have become necessary because of the far greater complexity of most eukaryotic genomes. Within the nucleus, RNA polymerase (pol) I is responsible for synthesizing most of the rRNA, pol II synthesizes mRNA and most of the snRNA, and pol III synthesizes a variety of small stable RNAs including tRNA, 5S rRNA and U6 snRNA. The templates transcribed by pols I, II and III are often referred to as class I, II and III genes, respectively.

The presence of three distinct nuclear RNA polymerases in eukaryotic cell extracts was originally identified because of their differential elution from columns carrying the ion-exchange resin DEAE–Sephadex. In addition to their distinct chromatographic properties and template preferences, they also display widely differing sensitivities to a toxin called α-amanitin, a cyclic octapeptide that is produced by the poisonous *Amanita* mushrooms. α-Amanitin works by interfering with the translocation process during RNA elongation. In mammals, pol II is the most sensitive to α-amanitin (50% inhibition at 25 ng/ml), whereas pol III displays intermediate sensitivity (50% inhibition at 20 μg/ml) and pol I is completely resistant. This property has been regularly exploited to determine which RNA polymerase is responsible for transcribing a particular gene; for example, if expression of a novel gene is unimpaired by 2 μg/ml α-amanitin, but is blocked by 100 μg/ml α-amanitin, one can infer that the gene is a template for pol III.

Biochemical characterization of the nuclear RNA polymerases proved to be a difficult task. One reason for this is their relative scarcity; for example, only 3–4 mg of pure pol I are obtained from 300 g of *Saccharomyces cerevisiae*. The second reason is that these enzymes are extraordinarily complicated, pols I, II and III from *S. cerevisiae* having 14, 12 and 17 subunits, respectively, by current estimates. Several of the subunits were difficult to identify because they comigrate in polyacrylamide gels. Furthermore, it has often been uncertain whether a copurifying polypeptide is a genuine subunit, an associated regulator or merely a contaminant. Such issues have sometimes been addressed by testing whether antibodies directed against the polypeptide in question can influence the function of the RNA polymerase. A powerful complementary approach has been to make use of genetic analyses. Genes have now been cloned from *S. cerevisiae* which encode all of the subunits of pols I, II and III.

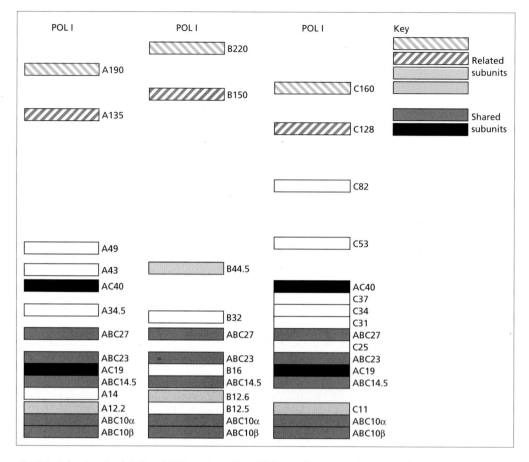

Fig. 2.1 Subunits of pols I, II and III from *S. cerevisiae*. Unique subunits are shown in white. Those shared by all three enzymes are shaded grey. The AC40 and AC19 subunits (in black) are found in both pol I and pol III; they are related to the pol II subunit B44.5. Also indicated is the similarity between the three largest subunits (A190, B220 and C160), the three second largest subunits (A135, B150 and C128) and between A12.2, B12.6 and C11.

This was achieved in several instances by screening expression libraries with antibodies against specific subunits. More commonly, DNA probes were generated on the basis of protein microsequence data derived from purified polypeptides. All of the subunit genes are single copy. Most of them have proved essential for viability, but yeast can survive without certain RNA polymerase subunits.

Because of the relative ease of genetic manipulation in yeast, the best characterized nuclear RNA polymerases are those of *S. cerevisiae*. Most of the information given in this chapter will pertain to this organism; however, the situation is thought to be very similar in higher eukaryotes. Pols I, II and III each consist of two large subunits and a complex array of smaller components that have sizes ranging from ~10 kDa to ~90 kDa (Fig. 2.1). The largest polypeptide of each is homologous to the largest polypeptide of the others and to the β′ subunit of prokaryotic RNA polymerases. Furthermore, the three second largest subunits are homologous to each other and to the prokaryotic β subunit. Pols I, II and III also share five common subunits, whilst two others are found in pols I and III, but not in pol II. In addition, each enzyme has its own set of unique polypeptides. The convention is to name a subunit with a letter and a number, such that the letter describes which polymerase(s) contains that subunit (with A for pol I, B for pol II and C for pol III) and the number indicates its size in kDa. For example, the 40-kDa subunit that forms part of pols I and III, but is not found in pol II, is referred to as AC40.

The catalytic activity of *Escherichia coli* RNA polymerase is provided by a heterotetramer composed of the β and β′ subunits and two copies of a smaller α subunit. A similar situation is thought to lie at the core of the nuclear RNA polymerases in eukaryotes. Thus, in each case the two largest subunits bear substantial homology to the β and β′ subunits, as mentioned above. In addition, pol II contains two copies of a subunit B44.5 that shows clear homology to the prokaryotic α subunits. Pols I and III instead carry one copy each of the AC40 and AC19 subunits, both of which display sequence similarity to the α polypeptide of *E. coli*; however, it is not clear that the α homology is significant in the case of AC19. It seems reasonable to assume that the β, β′ and α-homologous subunits constitute the functional core of pols I, II and III, performing the basic catalytic steps that are fundamental to transcription. This idea is supported by the fact that the largest and second largest subunits of each yeast RNA polymerase become affinity labelled during transcript synthesis by an ATP derivative that reacts with lysine residues. The same pairs of subunits can be photocrosslinked to short RNA molecules containing 4-thiouridine at the 3′ terminus. As well as nucleotides and RNA, they are also thought to interact with the DNA template. This is consistent with the fact that electron-microscopic studies place the two largest subunits in the vicinity of a deep cleft that has been shown to accommodate nucleic acid. It seems that the active site is formed by the concerted action of the two largest subunits. This probably involves multiple conserved domains within each polypeptide, and the order of these has been maintained through evolution (Fig. 2.2).

It is possible that additional subunits contribute to the active site in the eukaryotic enzymes. However, there is little evidence that this is the case. Indeed, progress has been slow in assigning functions to the majority of subunits. Comparison of database entries with the sequences of yeast pol subunits did not provide any informative relationships with different proteins, apart from subunits of other RNA polymerases. In a few cases, functional data has been obtained by testing the effects of antibodies in biochemical assays. For example, an antibody against C34 was found to strongly inhibit tRNA gene transcription by pol III *in vitro*, but had much less effect upon non-specific transcription of an artificial DNA polymer; this suggested that C34 is important for pol III to be directed to the appropriate templates, an idea which was

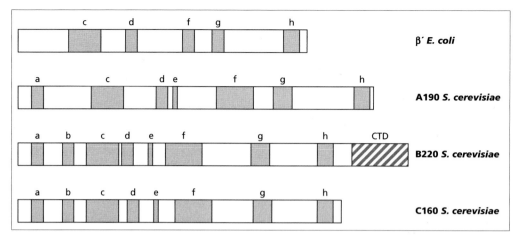

Fig. 2.2 Regions of homology between the largest subunits of pols I, II and III from *S. cerevisiae* and the RNA polymerase from *E. coli*. The blocks of homology are shown in black. Block a has been shown to bind zinc and has also been implicated in the interaction with DNA. Block g has been implicated in RNA binding. Blocks d and f have been implicated in transcript elongation. Also shown is the C-terminal domain (CTD) found in the largest subunit of pol II.

confirmed by subsequent studies. Thus, pol III is positioned at the initiation site of class III genes by protein–protein interactions between its C34 subunit and the transcription factor TFIIIB. It seems probable that many of the other unique subunits will also prove to be involved in class-specific interactions or functions. All of the subunits of pol III are required for viability in *S. cerevisiae*, but some of the pol I-specific and pol II-specific subunits are dispensable.

It is unclear what purpose is served by the five ABC subunits that are shared by all three pols. The absence of homologous subunits in prokaryotic RNA polymerases argues against a direct catalytic role in transcription. Nevertheless, they are each essential for viability in yeast and are highly conserved amongst eukaryotes, suggesting an important function in the transcription of nuclear genes. For example, the ABC10β subunit of *S. cerevisiae* is 70% identical to its human equivalent. Indeed, most of the ABC subunits from baker's yeast can be replaced by their human counterparts. Although homologues of ABC14.5 and ABC10β are only found in eukaryotes, archaebacterial and viral RNA polymerases contain subunits with homology to ABC10α, ABC23 and ABC27, implying a fundamental role for these. Indeed, biochemical analyses have shown that ABC23 is required for the structural and functional integrity of pol I in *S. cerevisiae*. Furthermore, electron microscopy has revealed that association of ABC23 induces a major conformational rearrangement of pol I. In the case of pol II, ABC23 has been shown by genetics to interact with B220. This subunit may therefore be regarded as a core component of the eukaryotic enzymes. There is very little known about the

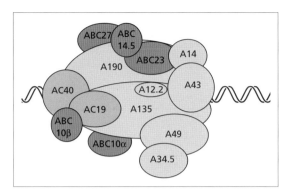

Fig. 2.3 Model of the subunit interactions that may occur in *S. cerevisiae* pol I. Biochemical and/or genetic evidence is available in support of most of the interactions shown in this model, although it must be regarded as tentative.

functions of the other common subunits, although ABC10α and ABC10β have been shown to bind to zinc.

Much more progress has been made in determining the interactions between the various subunits within the RNA polymerases. This information has been obtained from a combination of biochemical and genetic approaches. The two largest subunits are thought to interact extensively with each other and to be orientated in opposite directions. Zinc may be required for structural integrity, bringing together the N-terminus of the largest subunit and the C-terminus of the second largest, via their zinc-binding domains. The smaller subunits probably assemble around this central scaffold (Fig. 2.3).

X-ray analysis of eukaryotic RNA polymerases has been difficult because of their considerable size. However, the three-dimensional structures of pols I and II from *S. cerevisiae* have been studied by electron microscopy, after crystallization on charged lipid layers. X-ray diffraction has also been achieved for yeast pol II, providing structural information at 50 nm resolution. Both pols I and II have an irregular shape, with dimensions of approximately 15 × 11 × 11 nm. The most striking feature of these images is the presence of a groove or cleft 2.5 nm in diameter, which is wide enough to accommodate double helical nucleic acid. Electron crystallographic analysis of pol II in the act of transcribing has confirmed the presence of nucleic acid in the 2.5 nm cleft. A similar groove is observed in *E. coli* RNA polymerase, consistent with the high level of conservation between the two largest subunits of these enzymes. Furthermore, the very simple RNA polymerase of bacteriophage T7, which consists of a single subunit, also carries a comparable groove which has been shown to harbour the active centre. The DNA template is thought to associate with one face of yeast pol II, running up the 2.5 nm groove and bending ~90° during contact. This distortion is facilitated by the greater flexibility of single-stranded DNA, which forms as the duplex unwinds at the site of transcription. A mobile arm appears to close around the groove, which may be important to prevent the template from dissociating during transcription (Fig. 2.4).

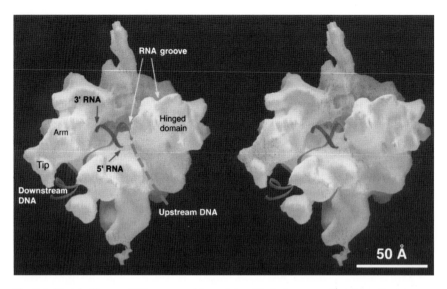

Fig. 2.4 Model of how pol II interacts with nucleic acids during transcript elongation. The DNA template and nascent RNA can be seen to pass through a groove or cleft in the polymerase surface. Sharp bending of the DNA is also indicated. The model is based on X-ray and electron-crystallographic data.

Given the strong phylogenetic conservation of their two largest subunits, it seems clear that pols I, II and III have evolved from a prokaryotic RNA polymerase. B220 and C160, the largest subunits of pols II and III, respectively, are more similar to each other than they are to A190, the largest of the pol I subunits. For example, homology block b is found in C160, B220 and in archaebacterial RNA polymerases, but is missing from A190 (Fig. 2.2). This has led to the suggestion that two nuclear RNA polymerases may have existed during one stage of evolution, one being the precursor of pol I and the other being the precursor of pols II and III. However, if this was the case it is paradoxical that pol III shares more subunits with pol I than it does with pol II.

It is curious that pol II has the fewest subunits, since it clearly transcribes a much larger and more diverse array of genetic templates than pols I or III. By contrast, pol III is the biggest and most complex of the enzymes, even though it is only required to express a modest collection of very short genes. It will be difficult to explain this anomaly until functions have been assigned to the various subunits. A partial explanation, however, may be that some of the functions performed by the additional subunits of pols I and III seem to have been adopted in the class II system by separate proteins that are not part of the pol II enzyme. For example, a number of subsidiary factors play important roles in helping pol II to elongate the nascent transcript. Some of these elongation factors will be described in Chapter 4. Without them, pol II terminates transcription prematurely, and would probably never reach the end of long templates such as the human dystrophin gene, which has 79 exons spanning

2.3 million base pairs. Pol I is also thought to use elongation factors, but none have been identified in the class III system. It is undoubtedly the case that pol III has much less need for such factors, because its templates are almost all shorter than 200 bp in length. Nevertheless, the C11 subunit of pol III from *S. cerevisiae* displays clear similarity, at both the sequence and functional levels, to an elongation factor called TFIIS that is used by pol II. Given this precedent, it may well be that the extra subunits in pols I and III contribute functions that are performed in the class II system by auxiliary factors that are separate from pol II. Indeed, it is certainly the case that pol II employs a far greater array of transcription factors than either of its cousins. Many of these are carried around with it in a large complex called the mediator. This associates with a unique feature of pol II, the C-terminal domain (CTD) of its largest subunit. The importance of the mediator and the various roles of the CTD will be described in subsequent chapters.

Further reading

Reviews

Asturias, F. J. & Kornberg, R. D. (1999) Protein crystallization on lipid layers and structure determination of the RNA polymerase II transcription initiation complex. *J. Biol. Chem.* **274**: 6813–6816.

Carles, C. & Riva, M. (1998) Yeast RNA polymerase I subunits and genes. In: *Transcription of Ribosomal RNA Genes by Eukaryotic RNA Polymerase*, I (ed. M. R. Paule) pp. 9–38, Springer Verlag, Heidelberg.

Mooney, R. A. & Landick, R. (1999) RNA polymerase unveiled. *Cell* **98**: 687–690.

Sentenac, A. (1985) Eukaryotic RNA polymerases. *CRC Crit. Rev. Biochem.* **18**: 31–90.

Thuriaux, P. & Sentenac, A. (1992) Yeast nuclear RNA polymerases. In: *The Molecular and Cellular Biology of the Yeast* Saccharomyces: *Gene Expression*, vol. 2, pp. 1–48. Cold Spring Harbor Laboratory, Cold Spring Harbor, NY.

White, R. J. (1998) RNA polymerase III. *RNA Polymerase III Transcription*. Springer Verlag, Heidelberg.

Woychik, N. A. & Young, R. A. (1994) Exploring RNA polymerase II structure and function. *Transcription: Mechanisms and Regulation*. Raven Press, New York.

Young, R. A. (1991) RNA polymerase II. *Annu. Rev. Biochem.* **60**: 689–715.

Selected papers

Determining the function of individual pol subunits

Chedin, S., Riva, M., Schultz, P., Sentenac, A. & Carles, C. (1998) The RNA cleavage activity of RNA polymerase III is mediated by an essential TFIIS-like subunit and is important for transcription termination. *Genes Dev.* **12**: 3857–3871.

Huet, J., Riva, M., Sentenac, A. & Fromageot, P. (1985) Yeast RNA polymerase C and its subunits. Specific antibodies as structural and functional probes. *J. Biol. Chem.* **260**: 15304–15310.

Riva, M., Schaffner, A.R., Sentenac, A. *et al.* (1987) Active site labelling of the RNA polymerase A, B, and C from yeast. *J. Biol. Chem.* **262**: 14377–14380.

Thuillier, V., Brun, I., Sentenac, A. & Werner, M. (1996) Mutations in the alpha-amanitin conserved domain of the largest subunit of yeast RNA polymerase III affect pausing, RNA cleavage and transcriptional transitions. *EMBO J.* **15**: 618–629.

Werner, M., Hermann-Le Denmat, S., Treich, I., Sentenac, A. & Thuriaux, P. (1992) Effect of mutations in a zinc binding domain of yeast RNA polymerase C (III) on enzyme function and subunit association. *Mol. Cell. Biol.* **12**: 1087–1095.

Three-dimensional structure

Fu, J., Gnatt, A. L., Bushnell, D. A. *et al.* (1999) Yeast RNA polymerase II at 5 Å resolution. *Cell* **98**: 799–810.

Jensen, G. J., Meredith, G., Bushnell, D. A. & Kornberg, R. D. (1998) Structure of wild-type yeast RNA polymerase II and location of Rpb4 and Rpb7. *EMBO J.* **17**: 2353–2358.

Poglitsch, C. L., Meredith, G. D., Gnatt, A. L. *et al.* (1999) Electron crystal structure of an RNA polymerase II transcription elongation complex. *Cell* **98**: 791–798.

Schultz, P., Celia, H., Riva, M., Sentenac, A. & Oudet, P. (1993) Three-dimensional model of yeast RNA polymerase I determined by electron microscopy of two-dimensional crystals. *EMBO J.* **12**: 2601–2607.

Evolutionary conservation

Memet, S., Saurin, W. & Sentenac, A. (1988) RNA polymerases B and C are more closely related to each other than to RNA polymerase A. *J. Biol. Chem.* **263**: 10048–10051.

Shpakovski, G. V., Acker, J., Wintzerith, M., Lacroix, J.-F., Thuriaux, P. & Vigneron, M. (1995) Four subunits that are shared by the three classes of RNA polymerase are functionally interchangeable between *Homo sapiens* and *Saccharomyces cerevisiae*. *Mol. Cell. Biol.* **15**: 4702–4710.

Chapter 3: DNA Recognition by Transcription Factors

The DNA sequences that control gene transcription are recognized in eukaryotes not by the RNA polymerases, but by transcription factors. Not all factors have this capability, but many contain domains that allow them to bind with high affinity and selectivity to short regions of DNA that have a particular base sequence. Often there is some flexibility in the DNA sequence that is recognized, but certain key bases will be crucial for binding. Many different types of polypeptide structure can be employed to contact DNA. This chapter will describe some of the key examples.

The homeodomain

Many metazoan transcription factors make use of a highly conserved structure called a homeodomain in order to recognize DNA. The homeodomain is ~60 amino acid residues long and is encoded by a 180-bp DNA sequence called a homeobox. Homeoboxes were first identified in many of the genes that help control embryogenesis in *Drosophila*. Examples include the segmentation genes *fushi tarazu* and *engrailed*, the homeotic genes *Antennapedia* and *Ultrabithorax*, and the *bicoid* gene that determines anterior polarity. For this reason, it was initially thought that the homeobox might have some special significance for developmental control. However, subsequent studies found it to be present in a wide range of genes encoding transcription factors. Indeed, ~1% of all genes contain a homeobox in the nematode *Caenorhabditis elegans*.

The first clue as to the function of the homeodomain came from the observation that it contains a region with significant homology to the **h**elix-**t**urn-**h**elix (HTH) motif that is present in many prokaryotic transcription factors. A protein motif is a conserved substructure that cannot fold independently, in contrast to a domain, which is a structure with the autonomous capacity to fold correctly. The HTH motif had been shown to be responsible for sequence-specific DNA recognition within various prokaryotic factors. This raised the possibility that the homeodomain is also capable of binding DNA. Confirmation that this is indeed the case came first from studies of the *Drosophila* protein Engrailed, which is critical for determining the polarity of segmentation during fly embryogenesis. Immunoprecipitation experiments revealed that the homeodomain of Engrailed binds preferentially to certain DNA sequences in the promoter of the *engrailed* gene. It is not uncommon for transcription factors to bind to sequences that regulate their own genes, as this provides an opportunity for feedback control (see Chapter 9).

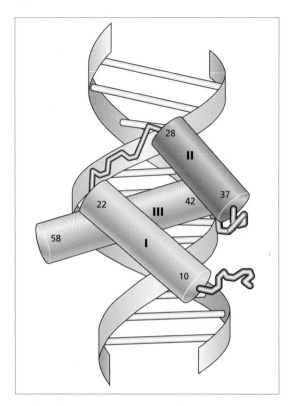

Fig. 3.1 Summary of the structure of the Engrailed homeodomain bound to DNA, as revealed by X-ray crystallography. Cylinders represent the three α-helices of the homeodomain, ribbons represent the sugar-phosphate backbone of the DNA and bars symbolize the base pairs. The recognition of helix (3) is shown in red.

The structure of the Engrailed homeodomain bound to DNA has now been visualized in great detail by means of X-ray crystallography. The homeodomain is globular and contains three α-helical regions of polypeptide (Fig. 3.1). Helices I and II lie almost exactly antiparallel to each other, whereas helix III crosses over these two at an angle of ~120°. As was predicted from sequence comparisons, helices II and III form a structure that closely resembles the prokaryotic HTH motif. One face of helix III is highly hydrophobic and is packed against helices I and II to form the hydrophobic interior of the protein domain. The other side of helix III is hydrophilic, with multiple positive charges; it fits into the major groove of DNA and makes extensive contacts with the bases and the sugar-phosphate backbone (Fig. 3.2). For this reason, helix III (homeodomain residues 42–58) is often referred to as the 'recognition helix'. Seven amino acids make positioning contacts with phosphate groups in the DNA backbone. These serve to align the recognition helix in the major groove so that base-specific contacts can be made by other residues. For example, hydrogen bonds are made with the bases by Gln50 and Asn51, whilst Ile47 makes hydrophobic contacts with a thymine (Fig. 3.3). The Engrailed homeodomain also has a flexible N-terminal arm that fits into the minor groove of the DNA, allowing residues 3, 5 and 7 to make base-specific contacts, whilst residue 6 interacts with the sugar-phosphate backbone (Figs 3.2 &

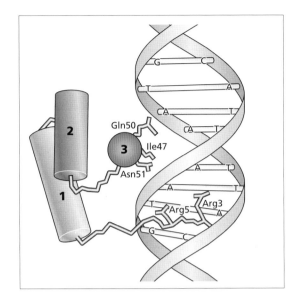

Fig. 3.2 Illustration of how the recognition helix and N-terminal arm of the Engrailed homeodomain interact with the major and minor grooves, respectively, of its DNA recognition site. Several critical contacts are also indicated.

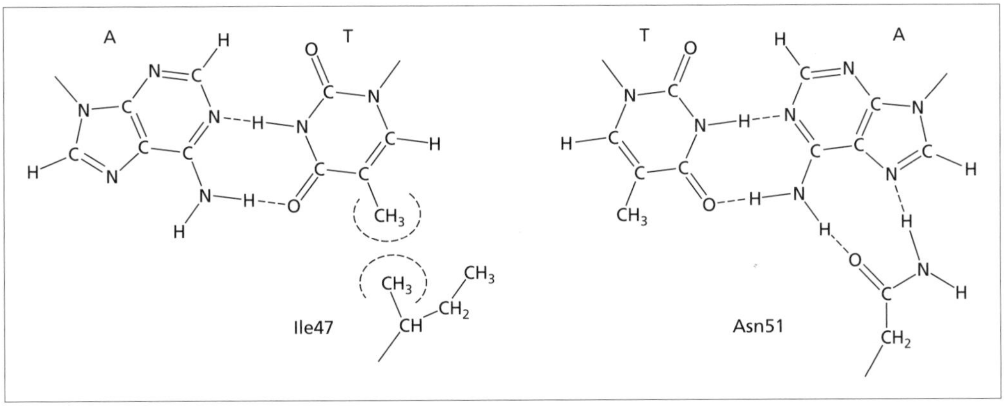

Fig. 3.3 Sketch illustrating the base-specific contacts made by Ile47 and Asn51 of the Engrailed homeodomain. Ile47 makes hydrophobic contacts whereas Asn51 makes use of hydrogen bonds.

3.4). Deletion of residues 1–6 from a homeodomain can reduce its DNA binding affinity by 10-fold. The N-terminal arm also makes an important contribution to the specificity of interaction.

Antennapaedia (Antp) is a *Drosophila* homeotic protein that is involved in determining cell identity during early embryogenesis. It also has a homeodomain and the structure of this has been determined by **n**uclear **m**agnetic **r**esonance (NMR). An advantage of NMR spectroscopy is that it can analyse dynamic structures in solution, whereas X-ray diffraction generally makes use of molecules that are held in a rigid crystal lattice. NMR analyses give much more emphasis to the dynamic features of protein–DNA interactions, with

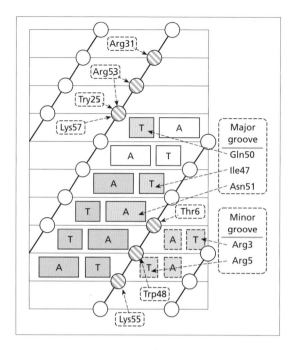

Fig. 3.4 Summary of all the contacts made between the Engrailed homeodomain and its recognition site, as revealed by X-ray crystallography.

amino acid side chains moving between two or more DNA contact sites on a time scale of milliseconds. This mobility resembles a continuous scanning process. Nevertheless, NMR showed that the structure of the Antp homeodomain bound to DNA in solution is extremely similar to that of the cocrystal of the Engrailed/DNA complex. In addition, NMR has also been used to study the solution structure of the Antp homeodomain in the absence of DNA. Comparison of the polypeptide structures observed in the presence and absence of DNA revealed that the Antp homeodomain does not undergo any major conformational change when it binds to its target site. The DNA also appears not to undergo any substantial rearrangement, although the width of the major groove increases slightly in order to accommodate helix III.

The homeodomain has been highly conserved through the course of evolution. For example, the human Hox-A7 homeodomain differs in only one out of 60 residues from that of Antp, its probable *Drosophila* homologue, despite the fact that vertebrates separated from insects more than 500 million years ago. An unexpected finding was that a yeast transcription factor called MATα2 has a very similar structure to the Antp homeodomain, even though these two proteins share only 28% sequence identity. Indeed, MATα2 not only displays the same protein fold, but uses a comparable docking arrangement with DNA and makes very similar base contacts. It is clear that this arrangement for positioning an α-helix in the major groove has been subject to very strong selective pressure.

Many homeodomain proteins recognize DNA sites with a very similar sequence, often involving ATTA (TAAT on the opposite strand). In the

Engrailed/DNA cocrystal, this consensus core sequence is contacted by Ile47 and Asn51. Ile47 is highly conserved amongst different homeodomains, whilst Asn51 is invariant. Whereas the common component of homeodomain recognition sites is contacted by these highly conserved residues, more variable amino acids in the recognition helix are primarily responsible for discriminating between individual DNA sites. Thus, Gln50 contacts the two C residues in the CCATTA sequence that is bound by Engrailed in the cocrystal. Gln50 is found at the same position in the homeodomains of the *Drosophila* proteins Fushi tarazu (Ftz) and Even-skipped, two transcription factors that also recognize the CCATTA site, as well as CAATTA. However, residue 50 is a Ser in Paired and a Lys in Bicoid, two *Drosophila* homeodomain proteins that cannot bind CAATTA. Paired becomes able to recognize CAATTA if its Ser50 is substituted for a Gln; by contrast, mutating adjacent residues in the Paired recognition helix to match those found in Ftz does not affect its binding specificity. Similarly, substituting the Ser50 of Paired into a Lys allows recognition of the Bicoid binding site. Bicoid binds preferentially to GGATTA sequences, for which Ftz has only low affinity. Changing Gln50 in the Ftz homeodomain into a Lys, like that of Bicoid, allows Ftz to bind with high affinity to GGATTA whilst reducing its affinity for CCATTA. These specificity swap experiments proved conclusively that residue 50 can play a decisive role in determining the DNA sequences that are recognized by a homeodomain. They also allowed the prediction that residue 50 contacts the two bases that precede the ATTA core sequence; this was confirmed subsequently by the X-ray and NMR structural studies.

An elegant genetic approach was used to test whether the same rules apply for homeodomain binding *in vivo*. This is an important issue, because the reaction conditions employed in a test tube are likely to be radically different from those occurring in a living cell. Three of the CCATTA sites that are bound preferentially by Ftz *in vitro* were substituted to GGATTA, for which it has low affinity. These sites were linked to a reporter gene and introduced into transgenic flies. Whereas wild-type Ftz activated expression from a reporter linked to CCATTA, no activation was obtained with the GGATTA construct. However, flies containing an engineered Ftz mutant, in which the recognition helix Gln50 had been substituted for Lys, were found to activate the reporter linked to GGATTA (Fig. 3.5). This demonstrates clearly that the rule deduced *in vitro* for DNA recognition by a homeodomain is also followed *in vivo*.

Despite the ability to manipulate the binding specificity of homeodomain proteins by substituting individual residues, there is no simple code relating which amino acid can contact any particular base. This depends on the precise geometry of the protein–DNA interface. Several different amino acid side chains can interact with a particular base if they are properly positioned. It is therefore not possible to use the polypeptide sequence of a DNA-binding domain to predict the base sequence of its recognition site, or vice versa. This is regrettable, since such a predictive capacity would be invaluable in characterizing novel transcription factors.

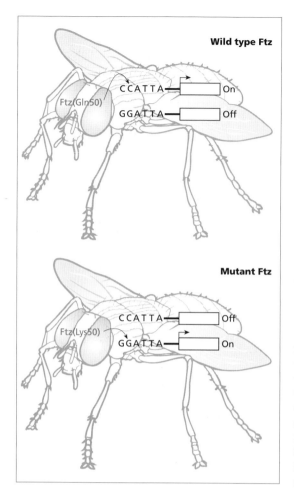

Fig. 3.5 Homeodomain residue 50 can dictate the specificity of DNA binding *in vivo*. In transgenic flies, wild-type Ftz activates transcription from CCATTA sites but not from GGATTA sites. Flies expressing a mutant form of Ftz, in which Gln50 has been substituted for Lys, activate transcription from GGATTA sites instead of the CCATTA sites. This shows that homeodomain residue 50 is an important determinant of DNA binding specificity *in vivo*.

The POU domain

A specialized version of the homeodomain was found within a structure called the POU domain, which was first identified as a long region of sequence similarity between the mammalian transcription factors Pit-1, Oct-1 and Oct-2 and a nematode factor called UNC-86. The term POU (pronounced pow) is derived from the names **P**it, **O**ct and **U**nc. POU domains have been found in many metazoan DNA-binding proteins, but they have not been recognized in plants or fungae, unlike homeodomains.

The POU domain is extremely unusual, because it contains two structurally independent domains that cooperate to function as a single DNA-binding unit. It consists of an ~75-amino-acid N-terminal POU-specific (POU$_S$) domain and a 60-residue C-terminal homeodomain (POU$_H$). These are joined by a linker polypeptide that varies considerably in both sequence and size (15–56

Fig. 3.6 Organization of the POU domain, which is composed of conserved POU_S and POU_H domains separated by a flexible linker of variable size and sequence.

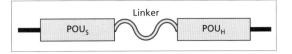

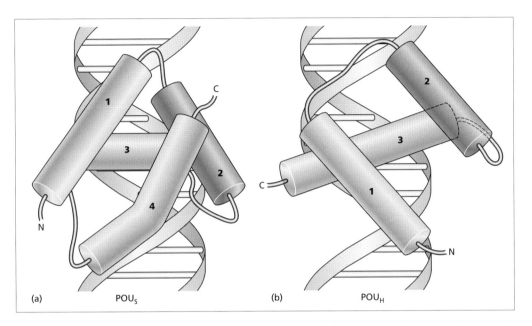

Fig. 3.7 The HTH motif in POU_S and POU_H domains. In both cases the HTH motif is formed by α-helices 2 and 3, which lie almost perpendicular to each other and are connected by a short turn. It is stabilized by two α-helices in the POU_S domain and by one in the POU_H domain. Helix 3 is the 'recognition helix' that makes many contacts with bases in the major groove of the DNA binding site.

residues) between different proteins (Fig. 3.6). Although their structures are autonomous, the POU_S domain and the POU_H variant of the homeodomain are always found together and must therefore have coevolved. X-ray crystallography has shown that the POU_H domain is nearly identical in structure to other homeodomains, such as Engrailed and Antp, and it is therefore regarded as a subfamily of the larger group. The POU_S domain is rather different, but like the homeodomains it contains an HTH motif. Whereas in homeodomains the HTH motif is stabilized by one additional α-helix, two are used in POU_S domains, giving four in total (Fig. 3.7). Curiously, the POU_S domain is found to resemble very closely the DNA-binding domains of repressor proteins encoded by the bacteriophages 434 and λ. The similarity is remarkable, given the vast evolutionary distance between vertebrates and phage, and the absence of POU domains in many lower organisms. It has been argued that the structure may constitute a particularly successful design for presenting the HTH motif.

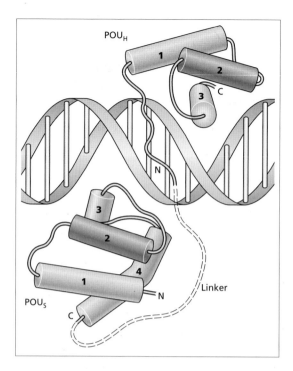

Fig. 3.8 Schematic diagram of the Oct-1 POU domain bound to its ATGCAAAT DNA site in the promoter of the histone H2B gene, as revealed by X-ray crystallography. The location of much of the linker region is unknown and is therefore indicated by a broken line.

Although the POU_S and POU_H domains can bind to DNA when separated, they do so with very low affinity. When joined together by the linker region, they can have very high affinity and specificity for particular sites. The X-ray crystal structures have been solved for Oct-1 and Pit-1 bound to their respect-ive DNA target sites. Oct-1 is found in most mammalian cell types; it binds to the sequence ATGCAAAT and activates transcription of a variety of genes, including those encoding histone H2B and various snRNAs. Crystallographic analysis showed that the Oct-1 POU_S and POU_H domains both lie in the major groove, but on opposite faces of the DNA (Fig. 3.8). The POU_S domain contacts the 5′ ATGC sequence, whereas the POU_H domain contacts the 3′ AAAT. It is notable that the latter resembles the TAAT core sequence that is recognized by many homeodomain proteins, including Engrailed and Antp.

The two POU subdomains of Oct-1 make no apparent contacts with each other when they are bound to DNA. They appear to be held together solely by the linker. One can think of the POU domain as a fused heterodimer, in which the dimer subunits, the POU_S and POU_H domains, are joined irreversibly through covalent attachment; however, most heterodimers require extensive contacts between the subunits, unlike the subdomains of Oct-1. Much of the linker was not visible by crystallography, which suggests that it is not a rigid structure. Because the length and sequence of the linker can vary greatly between different POU proteins, it is likely that there is substantial flexibility in how the subdomains are positioned on the DNA. Indeed, there is evidence that the sequences flanking the ATGCAAAT site can affect the conformation

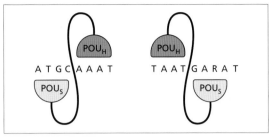

Fig. 3.9 The relative positions
of the POU$_S$ and POU$_H$
domains may be reversed
when Oct-1 binds to
ATGCAAAT and TAATGARAT
DNA sites.

of the Oct-1/DNA complex and thereby influence its ability to interact with other transcription factors. Furthermore, Oct-1 can also bind to the sequence TAATGARAT (R stands for purine), which is found in the promoters of several herpes simplex virus genes. Experiments involving the combined mutagenesis of both the POU domain and the binding site suggest that in this case the TAAT motif is bound by POU$_H$, whereas the GARAT is bound by POU$_S$. This places the POU$_S$ downstream of POU$_H$ on a TAATGARAT site, even though it is up-stream of it on an ATGCAAAT site (Fig. 3.9). A further example of the flexibility of the POU domain is provided by the neuronal transcription factor Brn-2. This protein can bind to DNA sites with spacings of either 0, 2 or 3 bp between its subdomain-binding sites. Furthermore, its POU$_S$ domain appears to be capable of switching orientations on different DNA elements. Such flexibility will clearly expand the repertoire of DNA sites that can be recognized by POU proteins.

Pit-1 is one of the founding members of the POU group and is expressed exclusively in the anterior pituitary gland. Its regulatory targets include the prolactin and growth hormone genes and it is required for the development of three of the five endocrine cell types. Many of the mutations which cause combined pituitary hormone deficiency in humans have been mapped to the POU domain of Pit-1. Patients with combined pituitary hormone deficiency suffer from a lack of prolactin, growth hormone and thyroid stimulating hor-mone, which often results in both mental and growth retardation. The linker between the POU$_S$ and POU$_H$ domains of Pit-1 is only 15 amino acids long, which makes it the shortest in any known POU protein. Because the linker is so short, the subdomains are unable to occupy opposite sides of the DNA, as occurs with Oct-1, and the POU$_S$ and POU$_H$ domains instead bind to perpen-dicular faces (Fig. 3.10).

The bZip domain

Whereas the POU domain is highly unusual because two independent DNA-binding domains are located in a single polypeptide, there are several examples in which two separate polypeptides are required to assemble a single functional DNA-binding domain. One example of this phenomenon is provided by the bZip domain, which consists of a **b**asic region followed by a

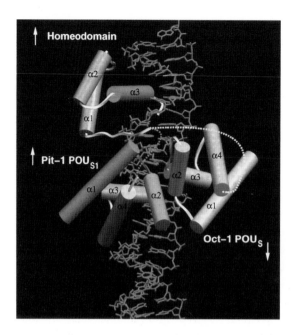

Fig. 3.10 Comparison of the relative positions of the Pit-1 and Oct-1 POU domains when bound to DNA, as revealed by X-ray crystallography. Compared with Oct-1, the POU$_S$ domain of Pit-1 is inverted in orientation and shifted to a different face of the DNA. Broken lines show the probable positions of the linker polypeptides, which are disorded in the crystal structures.

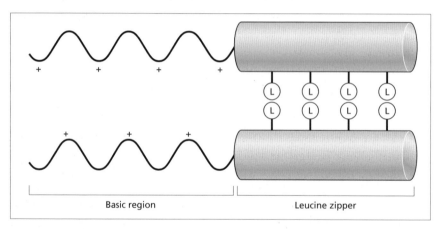

Fig. 3.11 Organization of a bZip domain. A basic region, which is responsible for contacting DNA, is followed by a leucine zipper that is responsible for dimerization. Two such structures must associate through the leucine zipper in order to assemble a functional DNA-binding domain.

structure called the leucine **zip**per (Fig. 3.11). The basic region is responsible for contacting DNA, but only after two such regions have been juxtaposed through dimerization mediated by the leucine zipper. This type of organization is found in many eukaryotic transcription factors, including GCN4 in yeast, and in mammals the **C**CAAT/**e**nhancer-**b**inding **p**rotein (C/EBP), the **c**AMP-**r**esponse **e**lement-**b**inding **p**rotein (CREB) and the transforming oncoproteins Fos and Jun that together make up the AP-1 family of factors.

The leucine zipper is ~30–40 amino acids long, with a leucine at every seventh residue to form a 'heptad repeat'. Site-directed mutagenesis shows that these leucines are very important for dimerization. The zipper region forms an amphipathic α-helix with the leucines and other hydrophobic residues arranged along one surface. NMR and X-ray crystallographic studies of a GCN4 homodimer and a Fos-Jun heterodimer have shown that the two zippers of a pair wrap around each other to give a parallel, two-stranded α-helical coiled-coil with a continuous hydrophobic interface. Although it was originally postulated that the leucine side chains would interdigitate like a zip, structural analyses revealed that they are in fact arranged in register; nevertheless, the colourful name has been retained. Most of the dimerization energy comes from hydrophobic interactions. However, hydrogen bonds and ion pairs can also occur and these provide selectivity to the pairing process. For example, although Jun can homodimerize, Fos cannot and it must instead heterodimerize with Jun in order to function. The reason for this specificity is that the interaction surface of Fos contains three negatively charged glutamate residues, which serve to repel another Fos molecule and thereby prevent heterodimerization. It has been possible to make Fos mutants that can homodimerize by substituting one of these glutamates for a positively charged lysine. A common way to determine whether two proteins interact stably with each other is by coimmunoprecipitation; this technique is described in Box 3.1.

Box 3.1 Coimmunoprecipitation assays

This technique is carried out to test if there is a stable interaction between two proteins, X and Y. An antibody against protein X is immobilized on insoluble agarose beads, often by chemical crosslinking. A solution containing proteins X and Y is then mixed with the beads and incubated for sufficient time to allow the antibody to bind to protein X. The beads are then pelleted by centrifugation and the supernatant is removed. Protein X will be found in the pellet, because it is held by the antibody and therefore immunoprecipitated. If protein Y binds stably to X, it will be found in the pellet as well, as a consequence of having been coimmunoprecipitated. If Y does not interact with X, it will remain in the supernatant. For example, an antibody against Fos will be expected to coimmunoprecipitate Jun, whereas an antibody against Engrailed should not.

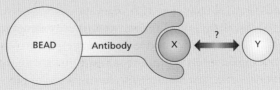

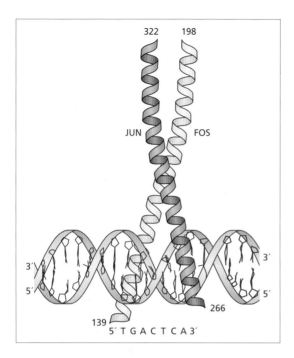

Fig. 3.12 Structure of the bZip domain of a Fos-Jun heterodimer bound to DNA, as revealed by X-ray crystallography. Numbers refer to amino acid residues in the Fos and Jun polypeptides. The DNA recognition sequence is shown underneath. Adapted with permission from *Nature* **373**. Copyright Macmillan Magazines Limited.

NMR has shown that the basic region is unstructured in solution, but becomes α-helical on binding to DNA; this can only occur once two such regions are brought together through dimerization. Once assembled, the bZip domain forms a Y-shaped structure, with the zippers forming the trunk and the two basic regions forming the arms of the Y. The basic α-helices then fit into the major groove of a 9–10-bp binding site and grip the DNA on opposite sides, making equivalent contacts with the bases (Fig. 3.12) These contacts are very similar for GCN4 and for Fos-Jun.

The bHLH domain

The bHLH domain contacts DNA by means of a basic region that is very similar to the basic regions of bZip proteins. However, instead of a leucine zipper, the **b**asic region is followed by a **h**elix-**l**oop-**h**elix dimerization motif in the bHLH domain. Examples of transcription factors that contain this structure include the myogenic factors MyoD and myogenin in vertebrates, the pair-rule gene product Hairy in *Drosophila* and the yeast protein Pho4 that regulates genes involved in phosphate metabolism. The HLH motif consists of two ~15-residue α-helical regions separated by a loop of 9–20 amino acids. The crystal structure of the MyoD bHLH domain homodimer bound to DNA shows that these two helices are at the dimer interface and form a parallel, left-handed, four-helix bundle. The basic region forms a continuous long α-helix with the first helix of the HLH motif (Fig. 3.13). The basic regions lie in the major groove contacting specific bases, just as in the Fos-Jun bZip complex.

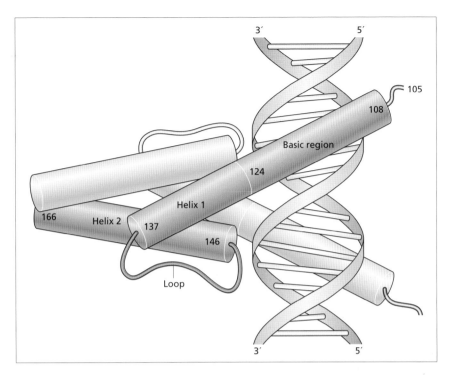

Fig. 3.13 Structure of the bHLH domain of a MyoD homodimer bound to DNA, as revealed by X-ray crystallography. Numbers refer to amino acid residues in the MyoD polypeptides.

The bHLH–Zip domain

Another group of transcription factors makes use of a bHLH domain that is followed immediately by a leucine zipper to give a bHLH–Zip domain. This arrangement provides a greater interface for dimerization, which may increase stability. It is found in the oncogene products Myc and Max, as well as many others. In this family, it is often found that neither the HLH nor the Zip motif alone are sufficient to mediate stable dimerization, but together they allow a much more efficient interaction. The distance between the HLH and the Zip can vary from one factor to another, and this serves to increase the specificity of dimerization because only certain combinations of monomers will have the correct spacing. The crystal structure of a Max homodimer bound to DNA revealed that each bHLH–Zip domain forms two long α-helices separated by a loop (Fig. 3.14) The N-terminal helix is 25 residues long and is formed from the basic region and the first helix of the HLH motif, whereas the C-terminal helix is 43 residues long and is composed of the second α-helix of the HLH motif and the leucine zipper region. As in Fos and Jun, the Zip regions form a parallel coiled-coil and the basic regions form α-helices that lie in the major groove perpendicular to each other on opposite faces of the DNA. Each

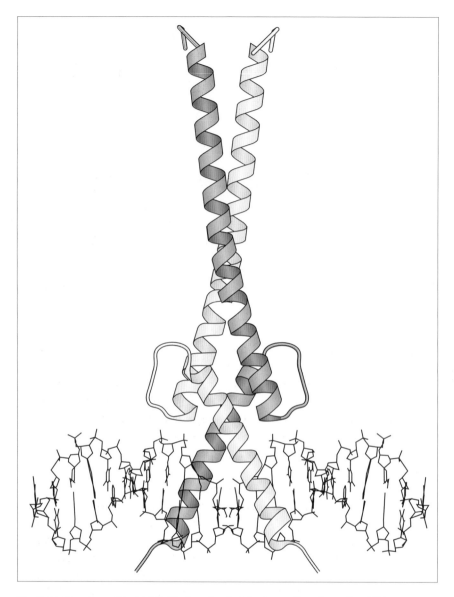

Fig. 3.14 Structure of the bHLH–Zip domain of a Max homodimer bound to DNA, as revealed by X-ray crystallography.

basic helix makes multiple contacts with the phosphodiester backbone and four sequence-specific contacts with the bases.

Zinc fingers

It is estimated that perhaps 1% of all mammalian proteins coordinate zinc, and this figure exceeds 3% in the nematode *C. elegans*. Although such proteins serve

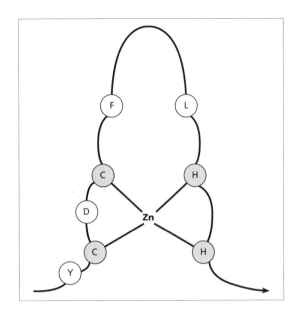

Fig. 3.15 Conserved residues that are found in the C$_2$–H$_2$ zinc fingers of *Xenopus* TFIIIA. A zinc ion is coordinated tetrahedrally by the two cysteine (C) and histidine (H) pairs. D, aspartate; F, phenylalanine; L, leucine; Y, tyrosine.

a range of functions, they include three important classes of DNA-binding protein. Each class has a very different structure, but in each case zinc serves as a flexible scaffold for supporting polypeptides. All three classes achieve DNA recognition using an exposed α-helix that fits into the major groove; the variation comes from the way that the helix is supported and presented to the DNA. The frequent exploitation of α-helices in this way, as also found in homeo-domains, bZip and bHLH proteins, reflects its compatibility with the size of the major groove and its intrinsic stability when exposed on a protein's surface.

The most widely occurring DNA-binding motif in eukaryotes is the C$_2$–H$_2$ zinc finger, which was first discovered in a transcription factor called TFIIIA that regulates the genes for 5S rRNA. TFIIIA is only 344 residues long, and most of its length consists of nine tandem repeats that each encode a zinc finger. It is very common for this class of zinc finger to occur in clusters. For example, four C$_2$–H$_2$ fingers are found in the *Drosophila* gap gene product Kruppel, three are found in the mammalian transcription factor Sp1 and the yeast factor SWI5, whilst a *Xenopus* protein called Xfin contains 37! A minimum of two adjacent C$_2$–H$_2$ fingers appears to be necessary for sequence-specific DNA recognition. Each finger is ~30 amino acids long and is charac-terized by pairs of cysteines and histidines that always occur at the same relative positions (Fig. 3.15). It forms a compact, globular domain in which a zinc ion is coordinated tetrahedrally by the cysteine and histidine residues. The zinc is essential for correct folding and for DNA binding. NMR and X-ray studies have shown that the C$_2$–H$_2$ finger is a compact globular domain com-posed of a 12-residue α-helix packed against an irregular β-sheet, with the zinc in between. The histidines lie on the inside face of the helix, whereas the

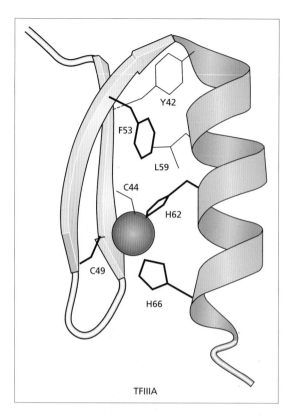

Fig. 3.16 Diagram of the fold of a C_2–H_2 zinc finger from *Xenopus* TFIIIA. The zinc ion is shown as a disc, the spiral represents an α-helix and the arrows represent β-sheets. Adapted with permission from *Nature* **356**. Copyright Macmillan Magazines Ltd.

cysteines flank a hairpin turn (Fig. 3.16). Cocrystal structures have been solved of several C_2–H_2 zinc finger proteins bound to DNA, including *Xenopus* TFIIIA and a *Drosophila* factor called Tramtrack that regulates transcription of the *ftz* gene. These analyses show that the α-helix of each finger lies in the major groove and contacts three base pairs (Fig. 3.17). The base contacts are almost entirely with one DNA strand. Positioning contacts to orientate the fingers are made between the sugar-phosphate backbone and several amino acids, including one of the zinc-coordinating histidines (Fig. 3.18). Adjacent fingers track around the major groove and contact consecutive units of 3 bp each. Flexible linkers between the fingers are set above the major groove and are not directly involved in contacting the DNA. This arrangement allows each finger to act as a separate module for sequence recognition (see Fig. 7.5).

A second class of DNA-binding domain that makes structural use of zinc is found in the nuclear receptors, a superfamily of ligand-activated transcription factors that mediate the response to extracellular stimuli such as steroid hormones, thyroid hormones, retinoic acid and vitamin D. The DNA-binding domains of these factors are 70–80 amino acids long and contain two zinc ions, each coordinated tetrahedrally with four cysteine residues. They are therefore sometimes referred to as C_2–C_2 zinc fingers. NMR and X-ray crystallography have revealed that these domains contain two similar loop-helix motifs that

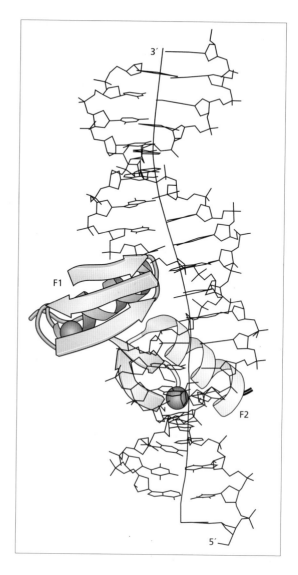

Fig. 3.17 Structure of the two C_2–H_2 zinc fingers of Tramtrack bound to DNA, as revealed by X-ray crystallography. The zinc ions are shown as discs, spirals represent α-helices and arrows represent β-sheets.

are folded together to form a single structural unit. Each motif has a zinc ion liganded by two cysteines at the start of the loop and two cysteines at the N-terminus of the α-helix (Fig. 3.19). The two α-helices in the domain are highly amphipathic and fold together so as to lie at nearly right-angles to each other; the hydrophobic surfaces on each pack together to produce a stable and extensive hydrophobic core that is essential to the structure of the domain. Cocrystal structures of the glucocorticoid receptor and the oestrogen receptor bound to DNA have shown that in both cases the first α-helix of the domain lies in the major groove, with amino acid side chains making hydrogen bonds with specific bases (Fig. 3.20). This is consistent with specificity swap muta-genesis experiments, which showed that discrimination between specific DNA

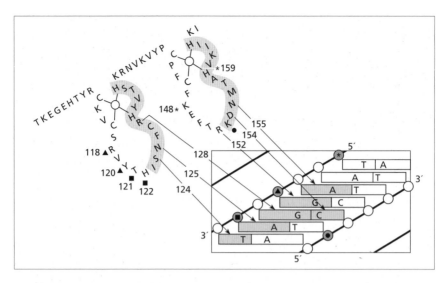

Fig. 3.18 Diagram illustrating the contacts made with DNA by the two C_2–H_2 fingers of Tramtrack. The amino acid sequence of the DNA-binding domain is written using the one-letter abbreviations. The residues incorporated into the α-helix are shaded. The zinc ions of the two fingers are represented as circles. Contacts between amino acids and bases are shown by arrows. Phosphate contacts are indicated by asterisks, squares, circles and triangles. Adapted with permission from *Nature* **336**. Copyright Macmillan Magazines Limited.

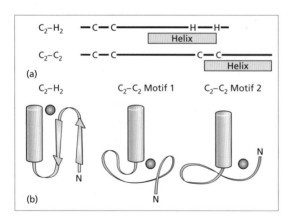

Fig. 3.19 Schematic comparison of the zinc-binding structures found in C_2–H_2 and C_2–C_2 fingers. (a) Comparison of the position of the α-helix relative to the zinc-coordinating residues in C_2–H_2 and C_2–C_2 fingers. (b) In C_2–H_2 fingers the zinc ion is held between the C-terminus of the α-helix (cylinder) and the β-sheet (arrow). In C_2–C_2 fingers the zinc is held between the N-terminus of the α-helix and the start of a loop; this motif is repeated.

sequences is achieved by three residues near the N-terminus of the helix of the first loop-helix motif. Ordered water molecules between certain amino acid side chains and the DNA increase the number of intermolecular hydrogen bonds and thereby improve the stability of the complex. The first loop and the second loop-helix of the domain make extensive positioning contacts to the phosphate backbone of the DNA.

Most members of the nuclear receptor superfamily bind DNA as dimers. They contain a dimerization domain that lies near the C-terminus of the protein (see Chapter 11). The receptors for thyroid hormone, vitamin D and

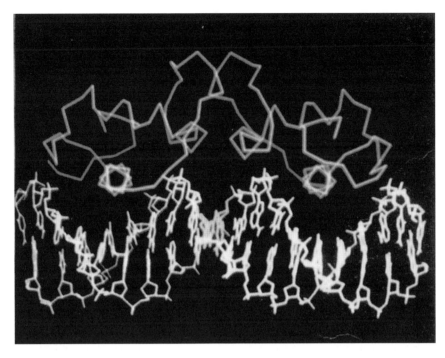

Fig. 3.20 Structure of the glucocorticoid receptor bound to DNA, as revealed by X-ray crystallography. A glucocorticoid receptor dimer sits above the DNA, with an α-helix from each monomer lying in adjacent major grooves. Adapted with permission from *Nature* **352**. Copyright Macmillan Magazines Limited.

retinoic acid function as heterodimers and can recognize DNA half sites with a range of spacings and orientations. By contrast, the steroid hormone receptors are much less flexible and bind only as homodimers to palindromic half sites separated by 3 bp and lying on the same face of the DNA duplex. The steroid receptors are much more constrained because they contain an additional dimerization interface within their DNA-binding domains. This maps to the loop of the second loop-helix motif, which lies on the surface of the protein and contacts the corresponding loop of another receptor molecule. Mutations that disrupt this interface are required to allow the glucocorticoid receptor to bind to sequences with altered half site spacings. Contact with the target DNA induces dimerization of the DNA-binding domains, resulting in highly cooperative interactions that determine the orientation of the recognition helices so that they bind in adjacent major grooves. By forming a cooperative dimer, the receptors can measure the spacing and orientation of the half sites, thereby increasing the specificity of binding considerably. Thyroid hormone receptors predominantly utilize the C-terminal dimerization domain, whereas the heterodimerization interface in the DNA-binding domain is weak; this gives them greater flexibility and allows them to bind to half-sites that are located on the opposite faces of the DNA.

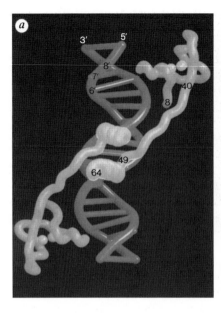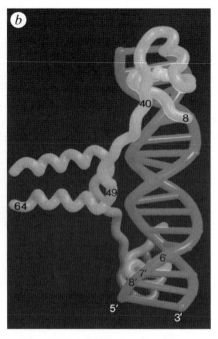

Fig. 3.21 Structure of Gal4 bound to DNA, as revealed by X-ray crystallography. The complex is viewed from two different angles. Adapted with permission from *Nature* **356**. Copyright Macmillan Magazines Limited.

A third class of DNA-binding domain that makes use of zinc is found in the transcription factor Gal4 that is involved in regulating sugar metabolism in *Saccharomyces*. The DNA-binding domain of Gal4 is found near its N-terminus (amino acids 1–74) and contains two closely spaced zinc ions that are coordinated tetrahedrally with six cysteine residues, two of which are shared between the metal ions. Once again, DNA recognition is achieved by an α-helix that fits into the major groove allowing amino acid side chains to make base-specific contacts. Gal4 binds as a dimer to a 17-bp palindromic sequence, with the half sites located on opposite faces of the DNA. The dimerization interface lies C-terminal to the DNA-binding domain and is joined to it by a polypeptide linker that tracks along the phosphate backbone, making contacts through a string of basic amino acids (Fig. 3.21). The dimerization elements meet at the dyad axis of the DNA site, above the minor groove; their helical chains lie parallel to each other, making contact through small hydrophobic amino acid side chains to form a short two-stranded coiled-coil.

p53

Yet another transcription factor that makes structural use of zinc is p53, an important tumour suppressor in humans. It has a large DNA-binding domain (~200 residues) with a unique structure. A β-sandwich, comprising

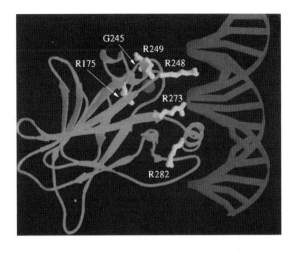

Fig. 3.22 Structure of the DNA-binding domain of p53 bound to DNA, as revealed by X-ray crystallography. Six residues are indicated that are hot spots for mutation in cancer. Of these Arg248 and Arg273 contact the DNA directly, whereas the others are necessary for the integrity of the structural elements at the interface with the DNA. Reproduced with permission from Cho, Complex Structure of a p53 tumour suspression DNA complex *Science* **265**. Copyright 1994, American Association for the Advancement of Sciences.

two antiparallel β-sheets, acts as a scaffold for a series of loop-based elements that form the interface with DNA. One component of these elements is a loop-sheet-helix that binds in the major groove where it makes base-specific contacts. The other is a pair of large loops that are held together, in part, by a tetrahedrally coordinated zinc atom; one of these loops presents a critical residue, Arg248, into the minor groove of the binding site (Fig. 3.22). The gene encoding p53 is the most frequent known site for mutations in cancer, with ~55% of solid human tumours carrying defects at this locus. The vast majority of substitutions are clustered in the DNA-binding domain and pre-vent p53 from regulating transcription. Six residues are particularly frequent targets for mutation, and these 'hot spots' all lie at or near the interface with DNA, far from the hydrophobic centre of the β-sandwich scaffold (Fig. 3.22). Arg248 and Arg273 are the most commonly mutated residues of all, and both these are responsible for contacting the DNA directly. The other hot-spot residues do not contact the DNA, but play critical roles in stabilizing the DNA-binding surface of the polypeptide. For example, Arg175 is involved in stabilizing interactions between the two loops that are bridged by the zinc atom. The fact that the hot-spot mutations will all disrupt the interaction with DNA, either directly or indirectly, provides a strong indication that DNA binding is central to the function of p53 as a tumour suppressor.

TATA-binding protein

Although α-helices are employed for DNA recognition by a large proportion of transcription factors, this is not invariably the case. An example of a factor with a very different way of interacting with DNA is the **T**ATA-**b**inding **p**ro-tein (TBP). The DNA-binding domain of TBP is 180 residues long and displays an astonishing level of evolutionary conservation. For example, the entire domain is identical between man and mouse and is 81% identical between

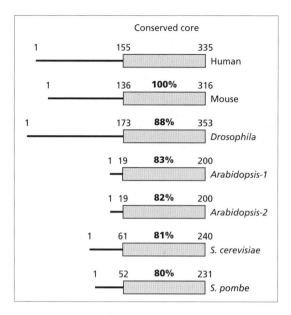

Fig. 3.23 Evolutionary conservation of the DNA-binding domain of TBP. The length of TBP varies through evolution, but the DNA-binding domain invariably constitutes the last 180 residues. This domain is shown coloured and its percentage identity with the corresponding human domain is indicated. The number of amino acids in each TBP is also shown. The plant *Arabadopsis* has two genes encoding TBP.

humans and the yeast *Saccharomyces cerevisiae* (Fig. 3.23). By contrast, the N-terminal region of the polypeptide varies through evolution in both its length and sequence. Not only is the C-terminal domain responsible for binding to DNA; but it also interacts with an extraordinary number of other polypeptides; this may have constrained it against change during the course of evolution.

X-ray crystallography has been used to examine the DNA-binding domain of TBP from humans, plants and yeast. In each case it adopts a highly symmetrical structure that resembles a saddle sitting astride the DNA (Fig. 3.24). It is composed of a pair of repeated halves that are only 31% identical in sequence, but are almost identical topologically, with the α-carbon atoms of the polypeptide chain following the same path through space in the two regions. The convex upper surface of the saddle consists of four α-helices, which make multiple protein–protein interactions with other transcription factors. The concave underside is responsible for binding to DNA, and this is formed from a curved, 10-stranded antiparallel β-sheet that is wide enough to accommodate the double helix. The DNA itself is severely distorted by TBP, being bent through an angle of ~80°. Its minor groove is widened and flattened, making extensive contact with the underside of TBP through many hydrophobic and some hydrophilic interactions. Localized untwisting of the TATA recognition sequence allows the bases to present a hydrophobic surface that is complimentary in shape to the underside of the saddle.

It is worth recalling that the TATA sequence bound by TBP resembles closely the core motif TAAT that is recognized by many homeodomain proteins. Despite the fact that these proteins are interacting with very similar DNA

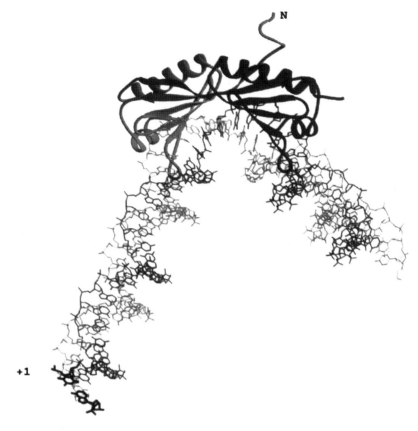

N

+1

Fig. 3.24 Schematic view of yeast TBP bound to DNA. The DNA is shown as an atomic stick model; the ~80° bend is apparent, as is the localized untwisting at the site of contact. TBP sits astride the DNA; its α-helices are shown as ribbons and its β-sheets are shown as arrows.

sequences, they nevertheless employ radically different structures in order to contact the bases. Like many other DNA-binding domains, homeodomains make use of an α-helix that fits into the major groove without distorting the DNA significantly. By contrast, TBP untwists and bends the DNA, to allow the bases to present a hydrophobic interface to a curved surface of antiparallel β-sheets. Clearly, it is possible to use very different molecular means to achieve similar ends.

Further reading

Reviews

Alber, T. (1992) Structure of the leucine zipper. *Curr. Opin. Genet. Dev.* **2**: 205–210.

Baxevanis, A. D. & Vinson, C. R. (1993) Interactions of coiled coils in transcription factors: where is the specificity? *Curr. Opin. Genet. Dev.* **3**: 278–285.

Freemont, P. (1993) Max in a complex affair. *Nature* **363**: 20–21.

Gehring, W. J., Qian, Y. Q., Billeter, M. *et al.* (1994) Homeodomain-DNA recognition. *Cell* **78**: 211–223.

Harrison, S. C. (1991) A structural taxonomy of DNA-binding domains. *Nature* **353**: 715–719.

Herr, W. & Cleary, M. A. (1995) The POU domain: versatility in transcriptional regulation by a flexible two-in-one DNA-binding domain. *Genes Dev.* **9**: 1679–1693.

Kerppola, T. & Curran, T. (1995) Zen and the art of Fos and Jun. *Nature* **373**: 199–200.

Klug, A. (1993) Opening the gateway. *Nature* **365**: 486–487.

Luisi, B. (1992) Zinc standard for economy. *Nature* **356**: 379–380.

Nelson, C. M. (1995) Structure and function of DNA-binding proteins. *Curr. Opin. Genet. Dev.* **5**: 180–189.

Prives, C. (1994) How loops, β sheets, and α helices help us to understand p53. *Cell* **78**: 543–546.

Schwabe, J. W. R. & Rhodes, D. (1991) Beyond zinc fingers: steroid hormone receptors have a novel structural motif for DNA recognition. *Trends Biochem. Sci.* **16**: 291–296.

Travers, A. A. & Schwabe, J. W. R. (1993) Spurring on transcription? *Curr. Biol.* **3**: 898–900.

Selected papers

The homeodomain

Kissinger, C. R., Liu, B., Martin-Blanco, E., Kornberg, T. B. & Pabo, C. O. (1990) Crystal structure of an engrailed homeodomain-DNA complex at 2.8 Å resolution: a framework for understanding homeodomain–DNA interactions. *Cell* **63**: 579–590.

Otting, G., Qian, Y. Q., Billeter, M. *et al.* (1990) Protein–DNA contacts in the structure of a homeodomain–DNA complex determined by nuclear magnetic resonance spectroscopy in solution. *EMBO J.* **9**: 3085–3092.

Treisman, J., Gonczy, P., Vashishtha, M., Harris, E. & Desplan, C. (1989) A single amino acid can determine the DNA binding specificity of homeodomain proteins. *Cell* **59**: 553–562.

The POU domain

Jacobson, E. M., Li, P., Leon-del-Rio, A., Rosenfeld, M. G. & Aggarwal, A. K. (1997) Structure of Pit-1 POU domain bound to DNA as a dimer: unexpected arrangement and flexibility. *Genes Dev.* **11**: 198–212.

Klemm, J. D., Rould, M. A., Aurora, R., Herr, W. & Pabo, C. O. (1994) Crystal structure of the Oct-1 POU domain bound to an octamer site: DNA recognition with tethered DNA-binding modules. *Cell* **77**: 21–32.

bZip, bHLH and bHLH-Zip domains

Ferre-D'Amare, A. R., Prendergast, G. C., Ziff, E. B. & Burley, S. K. (1993) Recognition by Max of its cognate DNA through a dimeric b/HLH/Z domain. *Nature* **363**: 38–45.

Glover, J. N. M. & Harrison, S. C. (1995) Crystal structure of the heterodimeric bZIP transcription factor c-Fos–c-Jun bound to DNA. *Nature* **373**: 257–261.

Landschulz, W. H., Johnson, P. F. & McKnight, S. L. (1988) The leucine zipper: a hypothetical structure common to a new class of DNA binding proteins. *Science* **240**: 1759–1764.

Ma, P. C. M., Rould, M. A., Weintraub, H. & Pabo, C. O. (1994) Crystal structure of MyoD bHLH domain–DNA complex: perspectives on DNA recognition and implications for transcriptional activation. *Cell* **77**: 451–459.

O'Shea, E. K., Klemm, J. D., Kim, P. S. & Alber, T. (1991) X-ray structure of the GCN4 leucine zipper, a two-stranded, parallel coiled coil. *Science* **254**: 539–544.

Zinc fingers

Fairall, L., Schwabe, J. W. R., Chapman, L., Finch, J. T. & Rhodes, D. (1993) The crystal structure of a two zinc-finger peptide reveals an extension to the rules for zinc-finger/DNA recognition. *Nature* **366**: 483–487.

Kraulis, P. J., Raine, A. R. C., Gadhavi, P. L. & Laue, E. D. (1992) Structure of the DNA-binding domain of zinc GAL4. *Nature* **356**: 448–450.

Luisi, B. F., Xu, W., Otwinoski, Z. *et al.* (1991) Crystallographic analysis of the interaction of the glucocorticoid receptor with DNA. *Nature* **352**: 497–505.

Marmorstein, R., Carey, M., Ptashne, M. & Harrison, S. C. (1992) DNA recognition by GAL4: structure of a protein–DNA complex. *Nature* **356**: 408–414.

Nolte, R. T., Conlin, R. M., Harrison, S. C. & Brown, R. S. (1998) Differing roles for zinc fingers in DNA recognition: structure of a six-finger transcription factor IIIA complex. *Proc. Natl Acad. Sci. USA* **95**: 2938–2943.

Schwabe, J. W. R., Chapman, L., Finch, J. T. & Rhodes, D. (1993) The crystal structure of the oestrogen receptor DNA-binding domain bound to DNA: how receptors discriminate between their response elements. *Cell* **75**: 567–578.

p53

Cho, Y., Gorina, S., Jeffrey, P. D. & Pavletich, N. P. (1994) Crystal structure of a p53 tumour suppressor–DNA complex: understanding tumorigenic mutations. *Science* **265**: 346–355.

TBP

Kim, J. L., Nikolov, D. B. & Burley, S. K. (1993) Co-crystal structure of TBP recognizing the minor groove of a TATA element. *Nature* **365**: 520–527.

Kim, Y., Geiger, J. H., Hahn, S. & Sigler, P. B. (1993) Crystal structure of a yeast TBP/ TATA-box complex. *Nature* **365**: 512–520.

Nikolov, D. B., Hu, S.-H., Lin, J. *et al.* (1992) Crystal structure of TFIID TATA-box binding protein. *Nature* **360**: 40–46.

Chapter 4: Basal Transcription by RNA Polymerase II

This chapter describes the minimal components that are involved in transcription by RNA polymerase (pol) II. It begins with a description of the important DNA elements of a pol II promoter and goes on to describe how proteins interact with the promoter core and how this can lead to the initiation of transcription. It then considers how the coding region of a class II gene is transcribed and what happens when the end of the gene is reached. The important proteins that are required for transcription by pol II are described here. These have been conserved through evolution, and organisms as diverse as mammals, flies and yeast use very similar sets of factors. Indeed, several homologues of eukaryotic transcription factors have been identified in archaebacteria. The effects of the regulatory sequences and factors will not be dealt with until the next chapter. Transcription in the absence of these regulatory components is referred to as basal; it may never occur *in vivo*, where additional stimulatory factors appear to be required, but it is helpful to begin with a simplified system in order to establish some of the fundamental principles. The next chapter will describe how basal transcription becomes activated in order to achieve physiological levels of expression.

Pol II promoters

Pol II transcribes the majority of genes. Eukaryotes contain between 5000 and 50 000 class II genes scattered throughout their genomes; this means that on average, one will occur every 200 000 base pairs in the most genetically complex organisms. The enormous variety of pol II templates is reflected in a diversity of promoter structure. The promoters are located around the start of each gene. They contain binding sites for various transcription factors, but the nature of these sites varies from one gene to another, as does their number and location. In most cases, several such sites will be found within a few hundred base pairs upstream of the transcription start site. However, factor recognition sites can also be located within the transcribed region or downstream of the termination sequence. Furthermore, important DNA elements that can influence the expression of class II genes are sometimes located many kilobases away; these remote regulatory regions are referred to as 'enhancers' (Fig. 4.1). The activity of enhancers is generally independent of orientation or distance from the start site, unlike the proximal promoter elements, which are often sensitive to their precise location. The DNA sequences that contribute to the control of class II genes are as diverse as the transcription factors that bind to them. The transcriptional machinery is able to gather and integrate this

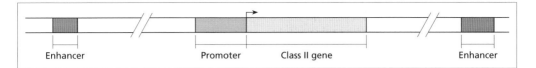

Fig. 4.1 The relative locations of promoter and enhancer elements in relation to a typical class II gene. The promoter generally extends for ~200 bp upstream from the transcription start site. The positions of enhancers is much more variable; they can be found several kilobases away either upstream or downstream from the coding region.

genetic information, interpreting it in distinct and often complex ways. This tremendous variety allows cells to tailor the expression of each gene to the particular circumstances under which its product is required.

Although the regulatory sequences of class II genes are so variable, two types of basal or core element are found at approximately the same position in a large proportion of cases. The best characterized of these is the TATA box, which is located ~25–30 base pairs upstream of the transcription start site in at least half of all pol II promoters. The TATA box has a consensus sequence TATAa/tAa/t, where the capital letters represent the most conserved bases and the small letters represent less consistent alternatives. The other common core element is the initiator, which is centred at the transcription start site. Initiators are rich in pyrimidines, but their precise sequence can vary considerably and they can be difficult to identify. Nevertheless, an initiator consensus has been derived as YYANa/tYY, where Y is a pyrimidine and N is any base; transcription begins at the A in this sequence. The defining feature of the core elements is that they are necessary and sufficient to allow accurately initiated transcription *in vitro*. Most pol II promoters have a TATA box or some kind of initiator. Many have both, and together these two elements can function synergistically to promote a high level of transcription. Some viruses, such as HIV and adenovirus, exploit this property to ensure that their genes are expressed very actively. Those promoters that do not contain a good match to a TATA box or an initiator are often transcribed inefficiently from multiple start sites, probably because polymerase recruitment is weak and imprecise.

Assembly of the preinitiation complex

TATA-directed assembly

The clearest picture of basal pol II transcription has been obtained using the **m**ajor **l**ate (ML) promoter of adenovirus. This template was chosen as the subject of many studies because it is more powerful than most cellular promoters and this makes it easier to work with. A reason it is so efficient is that it contains an initiator and a TATA box, and these both conform to the optimal consensus sequence. The principles that have been determined using this viral

Table 4.1 The basal pol II transcription initiation factors in humans.

Factor	Subunits
TFIID	Complex of ~700 kDa—binds TATA box—contains TBP (38 kDa) and ~10 TAFs—see Table 4.2
TFIIA	3 subunits—37 kDa, 19 kDa and 13 kDa
TFIIB	Single polypeptide of 33 kDa
TFIIF	2 subunits—30 kDa and 74 kDa—polymerase associated
TFIIE	Heterotetramer—two 34-kDa subunits and two 56-kDa subunits
TFIIH	Complex of 8–10 subunits—kinase and helicase activities—includes cdk7 and cyclin H (cdk-activating kinase, CAK) as well as ERCC2 and ERCC3 helicases involved in nucleotide excision repair (see Chapter 13)

template are believed to apply to many other promoters. However, no other example has been studied in such detail and significant variations may occur in different cases.

Assembly of a pol II preinitiation complex requires a number of distinct basal factors (Table 4.1). Chapter 3 described how the TATA-binding protein (TBP) interacts with its DNA recognition sequence. Binding of TBP to a TATA box is the first step in the assembly of the basal transcription apparatus on the adenovirus ML promoter (Fig. 4.2). TBP is a small polypeptide of 38 kDa in humans and only 27 kDa in yeast. As was described in Chapter 3, it is one of the most highly conserved proteins that we know (see Fig. 3.23). The C-terminal 180 amino acid residues constitute the DNA-binding domain and these are unchanged between man and mouse. Furthermore, 80% of them are identical when human TBP is compared to yeast. This remarkable evolutionary conservation indicates the extreme importance of TBP. The N-terminal region of TBP is considerably larger in animals than it is in plants or microorganisms; it is dispensable for most types of transcription, but plays a role in the synthesis of U snRNA.

Mammalian cells are thought to contain little or no free TBP. Instead, the protein is found in a number of complexes with polypeptides referred to as TAFs (**T**BP-**a**ssociated **f**actors). TAFs are named according to their size and the polymerase that utilizes them; thus, a 150-kDa TAF that is part of the pol II machinery is referred to as $TAF_{II}150$. The TBP-containing complex that binds to pol II promoters is called TFIID. This multisubunit conglomerate contains at least eight TAFs and has an aggregate mass of ~750 kDa (Table 4.2). The backbone of TFIID is thought to be provided by its largest subunit, $TAF_{II}250$, which serves as a scaffold along which other components assemble. TBP and the various TAFs make multiple protein–protein contacts within the TFIID complex; these interactions provide the structure with such stability that it can withstand exposure to 6 molar urea, a chaotropic reagent that would completely denature most proteins. Because of its size, TFIID can cover a significant region of DNA besides its TATA box recognition site. On some promoters, $TAF_{II}150$

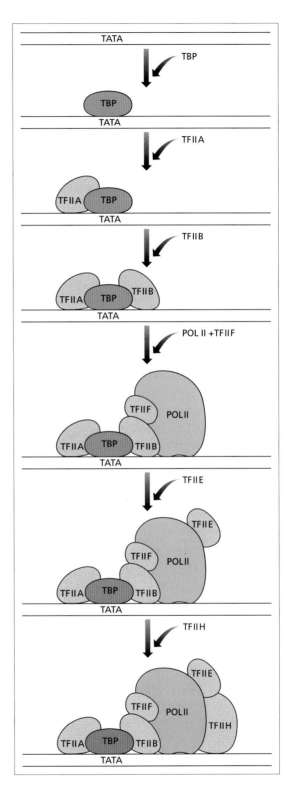

Fig. 4.2 Order of assembly of the basal pol II factors on the adenovirus ML promoter. Promoter recognition is achieved by binding of TBP to the TATA box. TBP will function as part of the TFIID complex, but the TAF subunits of TFIID have been omitted here for the sake of clarity. DNA-bound TBP recruits TFIIA and TFIIB. TFIIB, in turn, provides the docking site for a preformed complex of TFIIF bound to pol II. TFIIE and TFIIH are recruited last through interactions with pol II. Many of the steps in this assembly pathway may occur simultaneously through the recruitment of a preformed holoenzyme complex containing pol II and several basal factors.

Table 4.2 Subunits of human TFIID.

Subunit	Features
TBP (38 kDa)	Binds TATA box
TAF$_{II}$250	Scaffold; protein kinase and histone acetyltransferase activities
TAF$_{II}$150	Binds DNA downstream of TATA; loosely associated
TAF$_{II}$135	
TAF$_{II}$100	WD-40 repeats
TAF$_{II}$80	Similarity to histone H4
TAF$_{II}$55	
TAF$_{II}$31	Similarity to histone H3
TAF$_{II}$30	
TAF$_{II}$28	Atypical histone fold
TAF$_{II}$20	Similarity to histone H2b
TAF$_{II}$18	Atypical histone fold

provides subsidiary contacts with the DNA that increase the affinity of binding. In the case of the ML promoter, footprinting has shown that TFIID extends along the DNA from the TATA box to beyond the initiation site (Box 4.1 describes the technique of footprinting). Such an extended surface of interaction can improve the affinity and specificity of TFIID recruitment. This can be very significant, because association of TBP with a TATA box is both slow and inefficient. However, on some promoters the TAFs can actually destabilize the TBP/TATA interaction, thereby diminishing transcription. The TAF subunits of TFIID therefore play an important role in discriminating between different genes.

A factor called TFIIA binds directly to TBP and stabilizes its interaction with DNA, thereby raising the level of transcription. This may be of particular importance for promoters with weak non-consensus TATA sequences. Although TFIIA can be eliminated from highly purified *in vitro* systems, yeast cannot live without it. Chapter 3 described how TBP sits astride the DNA helix like a molecular saddle, with a stirrup hanging on each side. TFIIA contacts the N-terminal stirrup and the phosphodiester backbone immediately upstream of the TATA box, sitting on the opposite face of the DNA from TBP (Fig. 4.3). Together TFIIA and TBP make far more extensive interactions with DNA than either could alone, and this explains why TFIIA enhances the promoter-binding properties of TBP. Yeast TFIIA is composed of two subunits (32 and 13 kDa) encoded by a pair of genes that are both essential for viability. The two polypeptides embrace extensively along their lengths to form a boot-like structure that contains both a six-stranded β-barrel and a four-helix bundle. Human TFIIA is also encoded by a pair of genes, but the product of one is processed into two separate polypeptides so that human TFIIA consists of three subunits (37, 19 and 13 kDa).

When it binds to a TATA box, TBP distorts the DNA to generate a novel three-dimensional nucleoprotein structure to which other polypeptides can

Box 4.1 Footprinting

The DNase 1 footprinting assay measures the ability of a protein to protect a radiolabelled DNA fragment against digestion by DNase 1. DNase 1 is an endonuclease with little sequence specificity that will cut almost anywhere within a DNA molecule. Treatment of free DNA with just enough DNase 1 to cut on average once per molecule generates a population of fragments of all possible sizes that appear as a continuous ladder when resolved according to size on a denaturing gel. If a protein X binds the DNA at a particular site, then this protects that site against digestion by DNase 1; as a consequence, a gap or 'footprint' appears in the ladder of DNA molecules. The position of the footprint reveals where on the DNA molecule the protein X has bound. Chemical cleavage agents are sometimes used for footprinting instead of DNase 1.

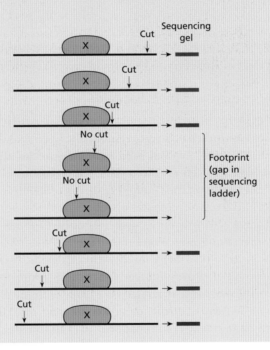

bind. This creates a binding site for TFIIB, a single polypeptide of 33 kDa in humans. TFIIB is essential for pol II transcription and is highly conserved among animal species (85% identity between humans and flies). However, yeast TFIIB is quite divergent, being only 35% identical with the human protein. The N-terminus of TFIIB forms a zinc-ribbon domain. Like TBP, TFIIB also contains a direct repeat, which allows it to adopt a symmetrical structure. The two repeats fold into compact globular domains composed of five α-helices and these are joined by a short random coil. The structure of these

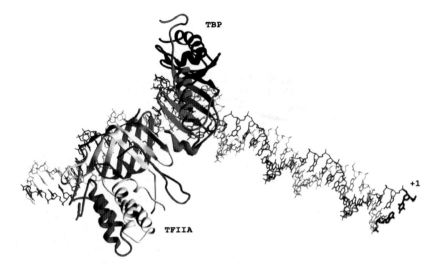

Fig. 4.3 Structure of a complex formed between yeast TBP, TFIIA and TATA-containing DNA, as revealed by X-ray crystallography. Proteins are represented as ribbon drawings, whereas DNA is shown as an atomic stick model, with the transcription start site indicated by +1. TBP binds to the TATA sequence. TFIIA interacts with the N-terminal repeat of TBP and contacts DNA upstream of the TATA box.

domains is similar to that of cyclins, a family of important proteins that activate the kinases which control cell cycle progression. The resemblance may suggest that cyclins and TFIIB have evolved from a common ancestor, but the same structure could have evolved on more than one occasion. The N-terminal repeat of TFIIB interacts with the C-terminal stirrup of TBP (Fig. 4.4). As well as binding TBP, TFIIB contacts promoter DNA both upstream and downstream of the TATA sequence. This is only possible because TBP bends its cognate site so violently that the DNA on either side of the TATA motif is brought into close proximity. A helix-turn-helix motif within its globular domain allows TFIIB to recognize specific bases located immediately upstream of the TATA box. By contrast, the downstream contacts are made with the phosphodiester backbone and are sequence-independent. Recruitment of TFIIB appears to be accompanied by dramatic conformational rearrangements, such that its repeated globular domains rotate through 90° with respect to each other. The interactions between TBP and TFIIB are asymmetrical, which partly explains why transcription only occurs in one direction from the TATA box.

TFIIB provides the platform for recruitment of pol II. The polymerase binds directly to the globular domain of TFIIB and this serves to position its active site over the transcriptional start site. As such, TFIIB can be regarded as a molecular bridge or ruler that determines how far from the TATA box initiation occurs. In most species, transcription begins approximately 25 bp downstream from the TATA sequence. However, in budding yeast the separation is typically 90 bp. This unusual property reflects a peculiarity of budding yeast TFIIB, as

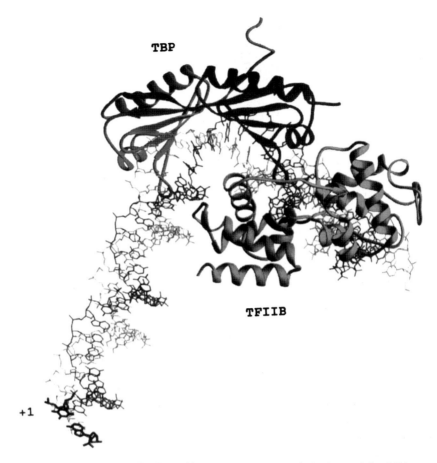

TBP

TFIIB

+1

Fig. 4.4 Structure of a complex formed between TBP, TFIIB and TATA-containing DNA, as revealed by X-ray crystallography. Proteins are represented as ribbon drawings, whereas DNA is shown as an atomic stick model, with the transcription start site indicated by +1. TBP binds to the TATA sequence. TFIIB interacts with the C-terminal repeat of TBP and contacts DNA both upstream and downstream of the TATA box.

shown by the fact that its substitution with TFIIB from a different species allows budding yeast to initiate 25 bp from the TATA box.

Pol II is tightly associated with a basal factor called TFIIF, even in the absence of DNA. TFIIF binds to the N-terminal region of TFIIB. Thus, the TFIIB platform provides separate docking sites for both pol II and TFIIF, which are recruited together as a complex (Fig. 4.5). This feature will undoubtedly improve both the affinity and specificity of recruitment, by increasing the surface of interaction. TFIIF is composed of 30-kDa and 74-kDa polypeptides, often referred to as RAP30 and RAP74 (for **R**NA **P**olymerase **A**ssociated). RAP30 has clearly evolved from a bacterial σ factor; not only does it display homology to these, but it can bind directly to *Escherichia coli* RNA polymerase. Furthermore, like *E. coli* σ^{70}, RAP30 has a cryptic DNA-binding domain that

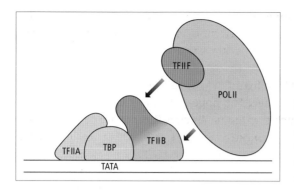

Fig. 4.5 Model illustrating how TFIIB provides separate docking sites for TFIIF and pol II, which are recruited as a preformed complex. TFIIF binds to the N-terminal region of TFIIB, whereas pol II binds to the central globular domain of TFIIB.

contacts promoter sequences. TFIIF suppresses inappropriate interactions between pol II and random DNA by inhibiting its association with sites outside promoters. As well as its major role in assembling the preinitiation complex, TFIIF also performs an important function in stimulating transcriptional elongation, as will be described later. Both the initiation and elongation activities of TFIIF can be increased by phosphorylation. Indeed, RAP74 is phosphorylated by a serine kinase activity intrinsic to $TAF_{II}250$, although it is unclear whether this is responsible for the stimulatory effect.

As mentioned in Chapter 2, the largest subunit of pol II has a C-terminal domain (CTD) that comprises multiple repeats of the heptapeptide sequence YSPTSPS. The number of times that this sequence is reiterated varies between species, with less complex organisms having fewer repeats; the yeast CTD has 26 copies and the human CTD has 52. These repeats are subject to phosphorylation in a reaction that is highly cooperative, so that the CTD is either unphosphorylated (form IIa) or heavily phosphorylated (form IIo). The two forms are quite discrete and intermediate levels of phosphorylation are rarely detected. The unphosphorylated CTD can bind directly to TBP and this may help anchor pol II at the start site prior to initiation; this interaction is lost when the CTD becomes phosphorylated. Only the unphosphorylated IIa form is assembled into preinitiation complexes, whereas actively transcribing polymerase is hyperphosphorylated; thus, CTD phosphorylation accompanies the transition from preinitiation to elongation.

Two more basal factors are recruited before transcription commences. The first is TFIIE, a tetramer composed of two 56-kDa subunits and two 34-kDa subunits. TFIIE binds directly to pol II and mediates the assembly of TFIIH. The latter is a complex factor, with nine different subunits and multiple enzymatic activities. The composition of TFIIH is highly conserved through evolution, and all of its subunits are essential for viability in yeast. Two of these subunits are helicases and can unwind the DNA duplex, one in each direction. Two others are cyclin H and **cyclin-dependent kinase** (cdk) 7, which together with a regulatory polypeptide form an activity called the **cdk activating kinase** (CAK). CAK is required to phosphorylate and activate the cdk family that plays a key role in regulating the cell cycle. It may be that evolution has selected CAK to perform two unrelated functions simply by chance; however, the involvement of this kinase in both transcription and cell cycle control may provide an opportunity for these processes to be coordinated. As will be described in Chapter 13, at least five of the subunits of TFIIH play an important role in the repair of DNA damage. This appears to allow the coupling of transcription to DNA repair, such that lesions in class II genes are corrected more rapidly than damage elsewhere in the genome. This is likely to be of considerable importance, as damage to an essential gene could be catastrophic for the cell. Defects in the subunits of TFIIH have been shown to cause three different types of human disease—xeroderma pigmentosum, Cockayne's syndrome and trichothiodystrophy. These conditions are associated with a broad range of symptoms, which can include profound growth retardation, delayed psychomotor development and brittle hair. Many of these effects are likely to result from malfunctions in transcription. Xeroderma pigmentosum is also characterized by an extreme sensitivity to sunlight and a high probability of skin cancer, which reflects the deficiency in DNA repair.

From the foregoing description, it will be appreciated that the fully assembled preinitiation complex is an extremely complicated machine. It contains at least 40 polypeptides and has a total molecular mass in excess of four megadaltons. As such, its complexity is comparable to that of a ribosome. TFIID and TFIIB are the only components that recognize specific bases, although other factors make sequence-independent contacts with the DNA. Multiple protein–protein interactions between the various basal factors serve to stabilize the preinitiation complex. TFIID has been shown to make direct contact with TFIIA, TFIIB, TFIIE, TFIIF and pol II. A remarkable number of different molecules can bind to TBP; it seems that each factor occupies only a restricted region of TBP and, in so doing, increases the surface area of the complex that is available for incoming polypeptides. The order of factor assembly (Fig. 4.2) was determined by a variety of complementary approaches, including kinetic analysis, template competition experiments, footprinting and band shift assays (Box 4.2 describes the band shift technique).

As already explained, most of this work made use of the adenovirus ML promoter. Although many of the principles that were determined for this model template are likely to be applicable generally, significant differences may also occur with other promoters. For example, it has been possible to reconstitute transcription of some class II genes using only subsets of the group of basal factors described here. An extreme case is provided by the immunoglobulin heavy chain gene, which can be transcribed *in vitro* using just TBP, TFIIB and pol II. It remains to be determined whether such a minimal system ever functions in living cells. However, in the archaebacterium *Sulfolobus shibatae*, homologues of TBP and TFIIB are sufficient to allow the RNA polymerase to transcribe a range of templates. This would suggest that the other basal factors that are found in eukaryotes have evolved more recently to allow additional scope for regulation.

Box 4.2 Band shift assay

The band shift assay tests the ability of a protein to bind a radiolabelled
DNA fragment as it migrates through a non-denaturing gel under the
influence of an electric current. Binding of the protein will reduce the
mobility of the DNA. If the fragment travels further in the absence of a
protein than it does in its presence, one can conclude that the protein
has affinity for that piece of DNA. Different proteins may retard a
fragment to different extents, depending on their size and shape. If two
proteins bind the same piece of DNA, they will reduce its mobility even
further. The band shift assay is very popular because it is easy, quick,
versatile and very sensitive. However, it does not reveal where a protein
is binding within a particular DNA fragment. Band shift assays are also
often referred to as gel shift, gel retardation, or electrophoretic mobility
shift (EMSA) assays.

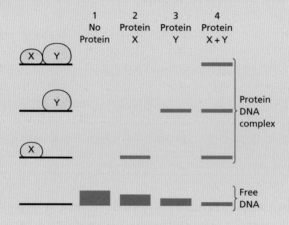

Initiator-directed assembly

Many pol II promoters lack TATA boxes. A well-characterized example is
provided by the murine **t**erminal **d**eoxynucleotidyl **t**ransferase (TdT) gene.
Despite lacking a TATA box, the TdT promoter requires all of the basal factors
that are utilized by the adenovirus ML promoter. It is particularly striking
that this TATA-less promoter should need TBP for its activity. However, point
mutations in the DNA-binding surface of TBP that prevent it from recognizing
TATA boxes and supporting transcription from the ML promoter do not impair
expression of the TdT gene; this indicates that TBP is brought to TATA-less
promoters in a manner that is independent of its sequence-specific DNA-
binding capability. One or more of the TFIID TAF subunits are involved in
recruiting TBP to genes that lack a TATA motif. Thus, whereas TBP can substitute
for the TFIID complex in supporting basal TATA-directed transcription, TFIID

Fig. 4.6 Organization of the promoter of the U2 snRNA gene. The principle promoter element is the PSE (proximal sequence element), which is located between 43 and 66 bp upstream of the initiation site. A distal sequence element (DSE) is found upstream of −214 and has similar properties to an enhancer. No TATA box or initiator element is present.

is necessary for the activity of TATA-less promoters. The TFIID subunits TAF$_{II}$250 and TAF$_{II}$150 can together recognize the initiator sequence. In addition, TAF$_{II}$150 alone can occupy the ML promoter for up to 40 bp downstream of the start site.

Several different proteins have been demonstrated to recognize initiator elements and may therefore participate in recruiting TFIID, presumably by binding to TAFs. One such factor is called TFII-I, which binds to some initiators and also to TBP. Another is named YY1, which has a high affinity for initiators containing the sequence CCAT. It may be that different initator-binding proteins function in different contexts. Once TFIID has been brought to the promoter through interactions with these factors, it is likely that the subsequent steps in assembling the preinitiation complex follow a similar route to that described for the ML promoter. Pol II itself has an intrinsic affinity for initiator sequences, and this is likely to assist recruitment in the presence of these elements.

PSE-directed assembly

The genes encoding snRNA, such as U1 and U2 that are required for mRNA splicing, have promoters that are quite different from those described above (Fig. 4.6). Most of these genes have neither a TATA box nor an initiator. Instead, assembly of the basal transcription machinery is directed by a **prox**imal **s**equence **e**lement (PSE) located between 46 and 66 bp upstream of the start site. The PSE is recognized by a factor called PTF (**P**SE-binding **T**ranscription **F**actor) or SNAPc (**sn**RNA **A**ctivating **P**rotein **c**omplex). This factor consists of five tightly associated subunits of 190, 50, 45, 43 and 19 kDa. The 190- and 50-kDa subunits interact with DNA at the PSE, whilst the 45- and 43-kDa subunits have affinity for TBP. It is therefore assumed that PTF can nucleate initiation complex assembly on snRNA genes by binding to the PSE and then using protein–protein interactions to recruit TBP.

The pol II holoenzyme

The description above of preinitiation complex assembly has assumed that the basal factors exist apart from one another in solution. However, there is

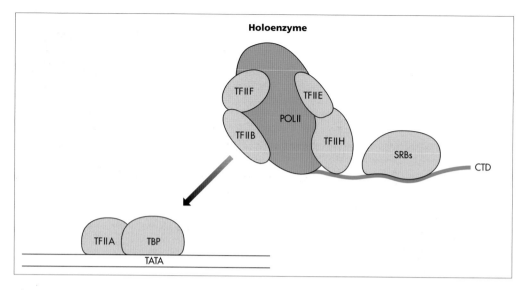

Fig. 4.7 Many components of the class II basal transcription apparatus may be recruited to promoters as a preformed holoenzyme complex containing TFIIF, TFIIB, TFIIE, TFIIH and SRB proteins associated with the CTD of pol II.

considerable evidence to suggest that some of these factors can interact with one another off the DNA and may therefore be recruited to promoters as preformed complexes. Several laboratories have identified complexes called 'holoenzymes', which contain hypophosphorylated pol IIa and subsets of the basal initiation factors. The precise composition of these complexes varies according to the source and the method of isolation. TFIIB, TFIIE, TFIIF and TFIIH have all been detected in some cases, but only TFIIF is invariably present. Holoenzymes also contain a set of nine proteins called SRB (for **S**uppressor of **R**NA polymerase **B**—RNA polymerases A, B and C are alternative and much less common names for RNA polymerases I, II and III, respectively). The SRB proteins bind to the unphosphorylated CTD and may function as a regulatory 'glue' to stabilize interactions between pol II and the basal factors. By recruiting a preformed holoenzyme to a promoter, a cell may bypass some of the stages involved in the stepwise assembly of a preinitiation complex (Fig. 4.7). This is likely to accelerate the process of transcription initiation. However, because each step in factor recruitment offers the potential for regulation, the binding of a holoenzyme may also decrease the flexibility of gene control.

Initiation

As soon as the preinitiation complex is assembled, transcription can begin. This involves three consecutive steps that follow in quick succession: promoter melting, in which the two strands of DNA around the start site are separated;

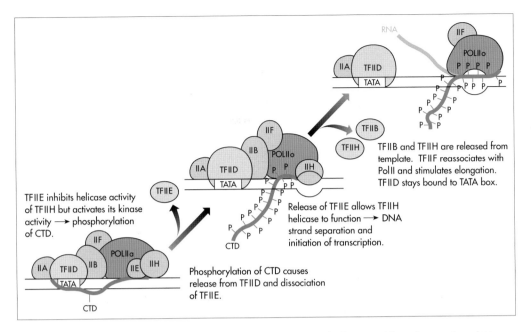

Fig. 4.8 Model of promoter clearance, illustrating the key transitions that are thought to accompany transcript initiation by pol II. In the preinitiation complex, TFIIE suppresses the helicase activities of TFIIH but stimulates its kinase to phosphorylate the CTD of pol II. Although the hypophosphorylated CTD binds to TBP, this interaction is broken once the CTD becomes hyperphosphorylated; this releases one of the constraints that anchors pol II to the promoter. TFIIE dissociates, which allows the helicase functions of TFIIH to unwind the DNA double helix in the vicinity of the initiation site, thereby providing a single-stranded template upon which pol II can synthesize RNA. As pol II moves away from the promoter, TFIIB, TFIIF and TFIIH are also released into solution.

initiation, when the first phosphodiester bond is formed in the nascent RNA; and clearance, in which pol II is released from the factors assembled at the promoter.

Promoter melting is necessary to allow the polymerase to gain access to the bases of the template strand. It is catalysed by TFIIE and TFIIH. This was shown by experiments in which the DNA was artificially premelted using short regions of mismatched sequence; TFIIE and TFIIH are no longer required if 6 bp of heteroduplex encompassing the start site are premelted in this way. One of the two helicase subunits of TFIIH is responsible for strand separation in an ATP-dependent step. This helicase activity is inhibited by TFIIE in the preinitiation complex. However, TFIIE also stimulates the kinase activity of TFIIH, causing it to phosphorylate the CTD of pol II. CTD phosphorylation induces conformational changes that release TFIIE from pol II and TFIIH. As TFIIE is displaced, the TFIIH helicase is no longer repressed and can unwind the DNA template, thereby allowing the polymerase to begin catalysing RNA synthesis (Fig. 4.8). Initially the region of melted DNA (a 'transcription bubble') becomes larger, but as the polymerase moves downstream the open region at the

Table 4.3 Some of the general pol II elongation factors.

Factor	Subunits	Activities
TFIIF	Heterodimer of RAP30 and RAP74	Suppresses pausing
Elongin (SIII)	Heterodimer of Elongin A (110 kDa), B (18 kDa), and C (15 kDa)	Suppresses pausing
ELL	80-kDa monomer	Suppresses pausing
TFIIS (SII)	38-kDa monomer	Prevents arrest and promotes nascent transcript cleavage

initiation site renatures into duplex. Passage of pol II requires release from contacts that would tether it to the start site. Most notably, the unphosphorylated CTD is bound to TBP which is anchored to the TATA box; phosphorylation by TFIIH dissociates this interaction and frees pol II to move along the gene. Promoter clearance is also accompanied by the release of TFIIB, TFIIE, TFIIF and TFIIH into solution. However, TFIIF can reassociate with the transcribing pol II and help with the elongation phase of RNA synthesis.

Elongation

Many protein coding genes are very large and can therefore take a considerable time to transcribe. An extreme example is the dystrophin gene, which is 2 Mb long and requires nearly 17 hours to be copied into a full-length transcript. It is therefore essential that the elongation phase of transcription is carried out efficiently; if pol IIo dissociated from its template or stalled at an appreciable rate, the longest RNA molecules might rarely be completed. In fact, a number of elongation factors have been identified that can facilitate transcript synthesis (Table 4.3). In the absence of such factors, purified pol II extends RNA chains at a rate of 100–300 nucleotides per minute; the process is interrupted regularly by pauses and sometimes arrests completely. *In vivo*, the elongation factors allow rates of up to 2000 nucleotides per minute to be achieved.

At least three different factors can stimulate mRNA synthesis by binding to pol II and suppressing transient pausing. One of these factors is TFIIF, which also plays an essential role in initiation, as already described. Different domains of TFIIF are responsible for the initiation and elongation functions. Another factor that helps protect pol II against pausing is elongin (also called SIII), which consists of a 110-kDa catalytic subunit (A) and two small regulatory subunits (B and C) that stimulate its activity. A third factor that reduces the duration of pausing is an ~80-kDa polypeptide called ELL. These three proteins are unrelated structurally but each binds the pol II enzyme. It is unclear how this facilitates elongation, but one possibility is that they keep the catalytic site of the polymerase correctly positioned with respect to the 3′ end of the growing RNA chain.

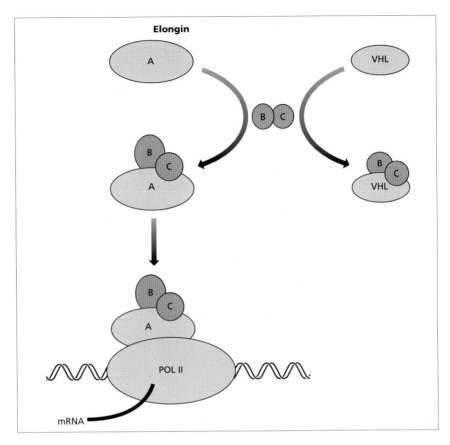

Fig. 4.9 Model for the regulation of elongin by the von Hippel–Lindau (VHL) tumour suppressor protein. A complex composed of the A, B and C subunits of elongin can bind to pol II and stimulate its ability to elongate nascent transcripts. VHL can bind to the B and C subunits of elongin and sequester them in a complex that is unable to assemble with elongin A.

The human gene encoding ELL is frequently disrupted by chromosomal translocations in acute myeloid leukaemias. This suggests that a malfunctioning elongation factor can contribute to oncogenesis. Support for this possibility is provided by evidence that elongin may be regulated by the **v**on **H**ippel–**L**indau (VHL) tumour suppressor gene. This gene is mutated in VHL disease, a rare genetic disorder that predisposes sufferers to a variety of cancers, such as clear-cell renal carcinoma. The VHL gene product interacts tightly with the B and C subunits of elongin and prevents these from binding and activating elongin A *in vitro* (Fig. 4.9). Some mutations occurring in VHL disease prevent the VHL tumour suppressor protein from regulating elongin. These observations have suggested a model in which mutation of VHL results in a deregulated hyperactive elongin factor, which may trigger the inappropriate expression of genes that contribute to neoplastic transformation. However, the situation is complex, and there is evidence that VHL performs several alternative functions.

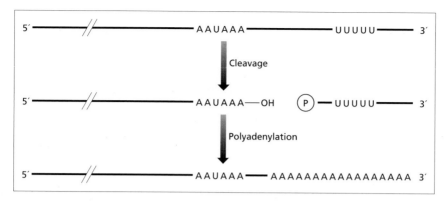

Fig. 4.10 Two steps are involved in the 3'-end processing of most pol II transcripts. The RNA is cleaved between the AAUAAA and the downstream U-rich element. This releases a downstream fragment with a 5' phosphate group that is rapidly degraded. The remaining upstream fragment is polyadenylated at its 3' hydroxyl group.

Despite the efforts of TFIIF, elongin and ELL, the progress of pol II sometimes becomes arrested. This can happen when the catalytic site of the polymerase loses contact with the 3' end of the growing RNA chain. A 38-kDa monomeric elongation factor called TFIIS (or SII) allows the arrested pol II to resume transcribing. It achieves this by binding pol II and stimulating an intrinsic endoribonuclease activity, which cleaves away those nucleotides of the nascent transcript that have become displaced from the catalytic site. This creates a new 3'-hydroxyl terminus for the RNA, which is positioned correctly to allow transcription to proceed. The ability of TFIIS to overcome arrest by promoting cleavage of the nascent RNA is of great importance. As a consequence, the gene encoding TFIIS is expressed ubiquitously and is highly conserved across species as diverse as archaebacteria, yeast and man. Thus, cells have evolved various elongation factors that perform complementary functions: TFIIF, elongin and ELL facilitate transcription by reducing the duration of pausing, whereas TFIIS reactivates pol II molecules that have arrested. Additional elongation factors have also been described.

Termination

The end of most mRNAs is marked by the sequence AAUAAA, often with a run of Us a little further downstream. Pol II continues to extend the transcript beyond this region, but the RNA is then cleaved between these two elements (Fig. 4.10). This process requires yet another array of factors (Table 4.4). First, the AAUAAA sequence is bound by the **C**leavage and **P**olyadenylation **S**pecificity **F**actor (CPSF), a complex of four subunits. The downstream U-rich region is bound by the heterotrimeric **C**leavage **St**imulation **F**actor (CstF). CPSF and CstF interact in a cooperative fashion to increase each other's affinity

Table 4.4 The mammalian factors required for 3'-cleavage and polyadenylation.

Factor	Subunits	Features
CPSF	160, 100, 73, 30 kDa	Binds AAUAAA; required for cleavage
CstF	77, 64, 50 kDa	Binds downstream element; required for cleavage
CF I	68, 59, 25 kDa	Binds RNA; required for cleavage
CF II	Uncharacterized	Required for cleavage
PAP	82 kDa	Required for cleavage; catalyses poly(A) synthesis
PAB II	33 kDa	Binds growing poly(A) tail; stimulates poly(A) elongation

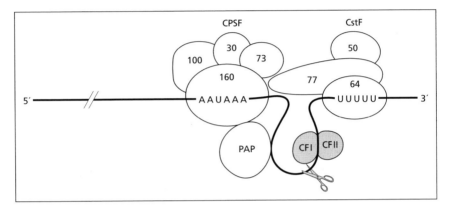

Fig. 4.11 Model of the complex responsible for RNA cleavage. CPSF binds to the AAUAAA sequence via its 160-kDa subunit. CstF binds to the U-rich region via its 64-kDa subunit and contacts CPSF with its 77-kDa subunit. The locations of CFI and CFII are uncertain. PAP binds to the 160-kDa subunit of CPSF, although it is uncertain whether this happens prior to RNA cleavage.

for RNA. Transcript cleavage also requires two poorly characterized **C**leavage **F**actors (CFI and II) and **p**oly(**A**) **p**olymerase (PAP), a single polypeptide with an ~200-residue serine/threonine-rich tail reminiscent of the CTD (Fig. 4.11). Immediately after chain scission, the upstream RNA is polyadenylated whereas the downstream RNA is degraded. CstF, CFI and CFII become dispensable once cleavage has occurred. PAP then adds adenylate to the 3' end of the transcript. The initial steps of polyadenylation are relatively slow. However, once a tail of at least 10 A residues has been formed, it is bound by a 33-kDa polypeptide called **P**oly(**A**)-**B**inding protein II (PABII) which accelerates catalysis. In the presence of PABII, polyadenylation is processive and rapid, with ~25 nucleotides added per second. CPSF and PABII tether the poly(A) polymerase to the transcript whilst it performs its function. Thus, polyadenylation is a highly cooperative process, as is the assembly of the cleavage complex. The reaction stops once the poly(A) tail has reached a length of ~250 nucleotides in mammals or ~80 nucleotides in yeast. It is unclear how the complex counts the number of A residues it has added, although this may be determined by

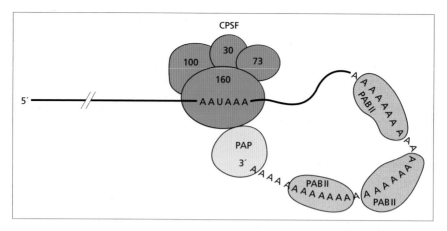

Fig. 4.12 Model of the polyadenylation complex. CPSF is believed to remain bound to the AAUAAA element during polyadenylation and to maintain its interaction with PAP. Multiple molecules of PABII bind to the extending poly(A) tail and stimulate its synthesis.

the stoichiometric binding of a defined number of PABII molecules as the tail extends (Fig. 4.12).

The procedure of 3′-end processing is an obligatory step in the synthesis of eukaryotic mRNA. Only once it is completed will the transcript leave the nucleus, although in many cases intronic sequences also need to be spliced out to create a mature message. The overwhelming majority of pol II transcripts lose a 3′ terminal non-coding fragment by endonucleolytic cleavage and then undergo polyadenylation through the steps described above. The only known cellular exceptions to this are the mRNAs encoding histones in animals. These use a distinct enzymatic apparatus to achieve endonucleolytic cleavage and receive no poly(A) tail. A number of viral mRNAs also provide exceptions to the general rule.

Further reading

Reviews

Aso, T., Conaway, J. W. & Conaway, R. C. (1995) The RNA polymerase II elongation complex. *FASEB J.* **9**: 1419–1428.

Burley, S. K. & Roeder, R. G. (1996) Biochemistry and structural biology of transcription factor IID (TFIID) *Annu. Rev. Biochem.* **65**: 769–799.

Colgan, D. F. & Manley, J. L. (1997) Mechanism and regulation of mRNA polyadenylation. *Genes Dev.* **11**: 2755–2766.

Conaway, J. W. & Conaway, R. C. (1999) Transcription elongation and human disease. *Annu. Rev. Biochem.* **68**: 301–319.

Drapkin, R. & Reinberg, D. (1994) The multifunctional TFIIH complex and transcriptional control. *Trends Biochem. Sci.* **19**: 504–508.

Goodrich, J. A. & Tjian, R. (1994) TBP–TAF complexes: selectivity factors for eukaryotic transcription. *Curr. Opin. Cell Biol.* **6**: 403–409.

Hahn, S. (1998) The role of TAFs in RNA polymerase II transcription. *Cell* **95**: 579–582.

Halle, J.-P. & Meisterernst, M. (1996) Gene expression: increasing evidence for a transcriptosome. *Trends Genet.* **12**: 161–163.

Hernandez, N. (1993) TBP, a universal eukaryotic transcription factor? *Genes Dev.* **7**: 1291–1308.

Koleske, A. J. & Young, R. A. (1995) The RNA polymerase II holoenzyme and its implications for gene regulation. *Trends Biochem. Sci.* **20**: 113–116.

Nikolov, D. B. & Burley, S. K. (1997) RNA polymerase II transcription initiation: a structural view. *Proc. Natl Acad. Sci. USA* **94**: 15–22.

Orphanides, G., Lagrange, T. & Reinberg, D. (1996) The general transcription factors of RNA polymerase II. *Genes Dev.* **10**: 2657–2683.

Reines, D., Conaway, J. W. & Conaway, R. C. (1996) The RNA polymerase II general elongation factors. *Trends Biochem. Sci.* **21**: 351–355.

Reines, D., Conaway, J. W. & Conaway, R. C. (1998) Mechanism and regulation of transcriptional elongation by RNA polymerase II. *Curr. Opin. Cell Biol.* **11**: 342–346.

Roeder, R. G. (1996) The role of general initiation factors in transcription by RNA polymerase II. *Trends Biochem. Sci.* **21**: 327–334.

Shilatifard, A., Conaway, J. W. & Conaway, R. C. (1997) Mechanism and regulation of transcriptional elongation and termination by RNA polymerase II. *Curr. Opin. Genet. Dev.* **7**: 199–204.

Svejstrup, J. Q., Vichi, P. & Egly, J.-M. (1996) The multiple roles of transcription/repair factor TFIIH. *Trends Biochem. Sci.* **21**: 346–350.

Tansey, W. P. & Herr, W. (1997) TAFs: guilt by association? *Cell* **88**: 729–732.

Uptain, S. M., Kane, C. M. & Chamberlin, M. J. (1997) Basic mechanisms of transcript elongation and its regulation. *Annu. Rev. Biochem.* **66**: 117–172.

Verrijzer, C. P. & Tjian, R. (1996) TAFs mediate transcriptional activation and promoter selectivity. *Trends Biochem. Sci.* **21**: 338–342.

Wahle, E. & Keller, W. (1996) The biochemistry of polyadenylation. *Trends Biochem. Sci.* **21**: 247–250.

Zawel, L. & Reinberg, D. (1995) Common themes in assembly and function of eukaryotic transcription factors. *Annu. Rev. Biochem.* **64**: 533–561.

Selected papers

Promoters

Aso, T., Conaway, J. W. & Conaway, R. C. (1994) Role of core promoter structure in assembly on the RNA polymerase II preinitiation complex. A common pathway for formation of preinitiation intermediates at many TATA and TATA-less promoters. *J. Biol. Chem.* **269**: 26575–26583.

Lagrange, T., Kapanidis, A. N., Tang, H., Reinberg, D. & Ebright, R. H. (1998) New core promoter element in RNA polymerase II-dependent transcription: sequence-specific DNA binding by transcription factor IIB. *Genes Dev.* **12**: 34–44.

Nakatani, Y., Horikoshi, M., Brenner, M. *et al.* (1990) A downstream initiation element required for efficient TATA box binding and *in vitro* function of TFIID. *Nature* **348**: 86–88.

Basal initiation factors

Akoulitchev, S., Mäkelä, T. P., Weinberg, R. A. & Reinberg, D. (1995) Requirement for TFIIH kinase activity in transcription by RNA polymerase II. *Nature* **377**: 557–560.

Liao, S., Zhang, J., Jeffrey, D. A. *et al.* (1995) A kinase–cyclin pair in the RNA polymerase II holoenzyme. *Nature* **374**: 193–196.

MacCracken, S. & Greenblatt, J. (1991) Related RNA polymerase-binding regions in human RAP30/74 and *Escherichia coli* σ70. *Science* **253**: 900–902.

Peterson, M. G., Inostroza, J., Maxon, M. E. *et al.* (1991) Structure and functional properties of human general transcription factor IIE. *Nature* **354**: 369–373.

Roy, R., Adamczewski, J. P., Seroz, T. *et al.* (1994) The MO15 cell cycle kinase is associated with the TFIIH transcription–DNA repair factor. *Cell* **79**: 1093–1101.

Serizawa, H., Mäkelä, T. P., Conaway, J. W., Conaway, R. C., Weinberg, R.A. & Young, R.A. (1995) Association of cdk-activating kinase subunit with transcription factor TFIIH. *Nature* **374**: 280–282.

Shiekhatter, R., Mermelstein, F., Fisher, R. P. *et al.* (1995) Cdk-activating kinase complex is a component of transcription factor TFIIH. *Nature* **374**: 283–287.

Sun, X., Ma, D., Sheldon, M., Yeung, K. & Reinberg, D. (1994) Reconstitution of human TFIIA activity from recombinant polypeptides: a role in TFIID-mediated transcription. *Genes Dev.* **8**: 2336–2348.

Transcription complex assembly

Buratowski, S., Hahn, S., Sharp, P. A. & Guarente, L. (1989) Five intermediate complexes in transcription initiation by RNA polymerase II. *Cell* **56**: 548–561.

Geiger, J. H., Hahn, S., Lee, S. & Sigler, P. B. (1996) Crystal structure of the yeast TFIIA/TBP/DNA complex. *Science* **272**: 830–836.

Ha, I., Roberts, S., Maldonado, E. *et al.* (1993) Multiple functional domains of transcription factor IIB: distinct interactions with two general transcription factors and RNA polymerase II. *Genes Dev.* **7**: 1021–1032.

Kim, J., Geiger, H., Hahn, S. & Sigler, P. B. (1993) Crystal structure of a yeast TBP/TATA-box complex. *Nature* **365**: 512–520.

Kim, J. L., Nikolov, D. B. & Burley, S. K. (1993) Co-crystal structures of TBP recognizing the minor groove of a TATA element. *Nature* **365**: 520–527.

Li, Y., Flnaan, P. M., Tschochner, H. & Kornberg, R. D. (1994) RNA polymerase II initiation factor interactions and transcription start site selection. *Science* **263**: 805–807.

Nikolov, D. B., Chen, H., Haley, E. D. *et al.* (1995) Crystal structure of a TFIIB–TBP–TATA-element ternary complex. *Nature* **377**: 119–128.

Nikolov, D. B., Hu, S.-H., Lin, J. *et al.* (1992) Crystal structure of TFIID TATA-box binding protein. *Nature* **360**: 40–46.

Ohkuma, Y. & Roeder, R. G. (1994) Regulation of TFIIH ATPase and kinase activities by TFIIE during active initiation complex formation. *Nature* **368**: 160–163.

Tan, S., Hunziker, Y., Sargent, D. F. & Richmond, T. J. (1996) Crystal structure of a yeast TFIIA/TBP/DNA complex. *Nature* **381**: 127–134.

Tsai, F. T. F. & Sigler, P. B. (2000) Structural basis of preinitiation complex assembly on human pol II promoters. *EMBO J.* **19**: 25–36.

Usheva, A., Maldonado, E., Goldring, H. *et al.* (1992) Specific interaction between the non phos-phorylated form of RNA polymerase II and the TATA-binding protein. *Cell* **69**; 871–881.

Verrijzer, C. P., Chen, J.-L., Yokomori, K. & Tjian, R. (1995) Binding of TAFs to core elements directs promoter selectivity by RNA polymerase II. *Cell* **81**: 1115–1125.

TATA-independent transcription

Carcamo, J., Buckbinder, L. & Reinberg, D. (1991) The initiator directs the assembly of a transcription factor TFIID-dependent transcription complex. *Proc. Natl Acad. Sci. USA* **88**: 8052–8046.

Chalkley, G. E. & Verrijzer, C. P. (1999) DNA binding site selection by RNA polymerase II TAFs: a TAF$_{II}$250-TAF$_{II}$150 complex recognizes the initiator. *EMBO J.* **18**: 4835–4845.

Kaufmann, J. & Smale, S. T. (1994) Direct recognition of initiator elements by a component of the transcription factor IID complex. *Genes Dev.* **8**: 821–829.

Martinez, E., Chiang, C. M., Ge, H. & Roeder, R. G. (1994) TAFs in TFIID function through the initiator to direct basal transcription from a TATA-less class II promoter. *EMBO J.* **13**: 3115–3126.

Pugh, B. F. & Tjian, R. (1991) Transcription from TATA-less promoter requires a multisubunit TFIID complex. *Genes Dev.* **5**: 1935–1945.

Roy, A. L., Malik, S., Meisterernst, M. & Roeder, R.G. (1993) An alternative pathway for transcription initiation involving TFII-I. *Nature* **365**: 355–359.

Usheva, A. & Shenk, T. (1994) TATA-binding protein independent initiation: YY1, TFIIB and RNA polymerase II direct basal transcription on supercoiled template DNA. *Cell* **76**: 1115–1121.

RNA polymerase II holoenzyme complex

Maldonado, E., Sheikhatter, R., Sheldon, M. *et al.* (1996) A human RNA polymerase II complex associated with SRB and DNA-repair proteins. *Nature* **381**: 86–89.

Ossipow, V., Tassan, J.-P., Nigg, E. A. & Schibler, U. (1995) A mammalian RNA polymerase II holoenzyme containing all components required for promoter-specific transcription initiation. *Cell* **83**: 137–146.

Thompson, C. M., Koleske, A. J., Chao, D. M. & Young, R. A. (1993) A multisubunit complex associated with the RNA polymerase II CTD and TATA-binding protein in yeast. *Cell* **73**: 1361–1375.

Initiation and promoter clearance

Goodrich, J. A. & Tjian, R. (1994) Transcription factors IIE and IIH and ATP hydrolysis direct promoter clearance by RNA polymerase II. *Cell* **77**: 145–156.

Lu, H., Zawel, L., Fisher, L., Egly, J.-M. & Reinberg, D. (1992) Human general transcription factor IIH phosphorylates the C-terminal domain of RNA polymerase II. *Nature* **358**: 641–645.

Maxon, M. E., Goodrich, J. A. & Tjian, R. T. (1994) Transcription factor IIE binds preferentially to RNA polymerase IIa and recruits TFIIH: a model for promoter clearance. *Genes Dev.* **8**: 515–524.

Wang, W., Carey, M. & Gralla, J. D. (1992) Polymerase II promoter activation: closed complex formation and ATP-driven start site opening. *Science* **255**: 450–453.

Elongation factors

Aso, T., Lane, W. S., Conaway, J. W. & Conaway, R. C. (1995) Elongin (SIII): a multisubunit regulator of elongation by RNA polymerase II. *Science* **269**: 1439–1443.

Bengal, E., Flores, O., Krauskopf, A., Reinberg, D. & Aloni, Y. (1991) Role of the mammalian transcription factors IIF, IIS, and IIX during elongation by RNA polymerase II. *Mol. Cell. Biol.* **11**: 1195–1206.

O'Brien, T., Hardin, S., Greenleaf, A. & Lis, J. T. (1994) Phosphorylation of RNA polymerase II C-terminal domain and transcriptional elongation. *Nature* **370**: 75–77.

Tan, S., Conaway, R. C. & Conaway, J. W. (1995) Dissection of transcription factor TFIIF functional domains required for initiation and elongation. *Proc. Natl Acad. Sci. USA* **92**: 6042–6046.

Reinitiation

Zawel, L., Kumar, K. P. & Reinberg, D. (1995) Recycling of the general transcription factors during RNA polymerase II transcription. *Genes Dev.* **9**: 1479–1490.

Chapter 5: Activating RNA Polymerase II Transcription

Intro

Basal pol II transcription is extremely inefficient. Although it has been studied extensively *in vitro*, it may never occur within a cell. Significant levels of expression of a pol II template require the intervention of stimulatory transcription factors that bind DNA at sites which are distinct from the TATA box and initiator regions where the basal factors assemble. Unlike the basal machinery, these stimulatory factors tend to be specific for particular genes and cell types. They are therefore able to confer unique expression patterns upon individual genes. Indeed, most genes display a regulated pattern of transcription during development or in response to changes in the cellular environment. The abundance of different mRNAs can vary by up to four orders of magnitude within a cell, primarily due to dramatic differences in the rates at which individual genes are transcribed. This chapter will address how basal rates of pol II transcription can be elevated to achieve physiological levels of expression.

Enhancers

The promoters used by pol II are typically located within about 200 bp of the transcription start site, whereas enhancers in higher organisms may be many kilobases away, either upstream or downstream. The ability of enhancers to function over such long distances provoked early speculation that their mode of action is radically different from that of promoters. However, it is now apparent that these two types of element are fundamentally similar. Each is composed of several discrete functional modules of DNA, which are the binding sites for individual transcription factors. The relative spacing and orientation of these modules can be extremely flexible. Some modules can be found in either a promoter or an enhancer context, for example the GC box which binds Sp1 or the octamer element that binds Oct-1 and Oct-2. Indeed, sometimes enhancers overlap with promoter sequences. It may perhaps be most accurate to speak of an 'enhancer effect', rather than a rigidly defined physical entity. Nevertheless, the term enhancer is generally applied to DNA elements that can stimulate transcription from a promoter when placed in either orientation, upstream or downstream and from distances exceeding a kilobase.

The specific expression pattern of a gene is ultimately governed by combinatorial interactions amongst the transcription factors that bind to the various regulatory modules within its promoter and enhancer(s). An enhancer located 1 kb upstream of the muscle creatine kinase gene provides a

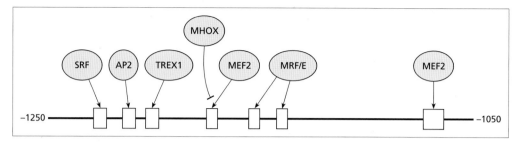

Fig. 5.1 Organization of an enhancer located ~1 kb upstream of the mouse muscle creatine kinase gene. The diagram indicates the identities and approximate sites of binding of various factors that are known to contribute to the activity of the enhancer in skeletal muscle cells.

good example of the potential complexity of these combinatorial interactions. This enhancer is less than 200 bp long but contains binding sites for at least six different transcription factors (Fig. 5.1). Variations in the abundance or activity of each of these factors can radically alter the function of the enhancer and thereby modulate transcription of the gene.

Many of the early insights into the characteristics of enhancers came from studies of a small DNA tumour virus called **S**imian **v**irus **40** (SV40). The genome of SV40 is a short circular DNA molecule of 5.2 kb that contains only two transcription units. These are transcribed in opposite orientations and have overlapping promoters and a shared enhancer (Fig. 5.2). The enhancer can stimulate transcription by over two orders of magnitude. The major contribution to this enhancer effect is provided by a tandem duplication of 72 bp. Pioneering experiments with this system revealed that it is possible to construct synthetic enhancers by isolating individual modules from the SV40 enhancer and multimerizing them. The artificial constructs are active if the modules are placed within 50 bp of each other. These experiments revealed that, at its simplest, an enhancer consists of a cluster of binding sites for transcriptional activator proteins. The clustering is important, as isolated modules have little activity; a pair of sites together produces much greater activation than the additive effect of the same two sites when separated, a property known as synergism. Unlike these synthetic constructs, real enhancers are composed of sites for more than one type of activator, as shown in Fig. 5.1. Some of the activators may only function in particular cell types, making the behaviour of the enhancer tissue-dependent. For example, the **m**yocyte **e**nhancer binding **f**actor 2 (MEF2) is most abundant in muscle, where it contributes to the cell type specificity of the muscle creatine kinase enhancer. However, some of the other factors that are important for the function of this enhancer are expressed in a broad range of cells, such as the **s**erum **r**esponse **f**actor (SRF). The enhancer of JCV, a virus which causes the demyelinating disease progressive multifocal leukoencephalopathy, exhibits a strong preference for fetal glial cells. By contrast, the SV40 enhancer can function in a wide

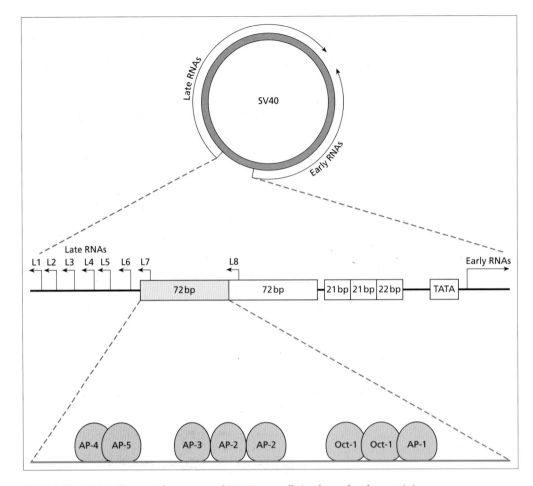

Fig. 5.2 The SV40 enhancer. The genome of SV40 is a small circular molecule containing early and late genes that are transcribed in opposite orientations (top). The SV40 enhancer is positioned just upstream of these (middle). The early RNAs are initiated 25–30 bp downstream of a TATA box. Their synthesis is stimulated by a series of three 21–22-bp promoter elements, each of which is bound by two molecules of the zinc finger transcription factor Sp1. Their synthesis is also activated strongly by the enhancer, which contains a direct repeat of 72 bp. Multiple transcription factors bind to these repeats, including Oct-1 and AP-1 (bottom). At least eight initiation sites give rise to the late RNAs and two of these lie within the 72-bp repeats of the enhancer. The origin of viral replication lies within the early transcribed region, just downstream from the Sp1 sites.

variety of tissues and hosts, ranging from humans to amphibia. The combined properties of the various activators together confer the specific characteristics of a particular enhancer. This combinatorial system provides the opportunity of achieving a huge diversity of enhancers out of a much smaller number of discrete modules. This constitutes a prime example of genetic economy. Furthermore, because of their flexibility, new enhancers can evolve simply by recombining the modules from pre-existing ones. The ability to generate

Box 5.1 Reporter genes

Many of the studies to investigate the function of promoters, enhancers and the factors which bind to them have made use of reporter genes. Three commonly used examples are the *lacZ* gene of *Escherichia coli*, the *E. coli* gene encoding **c**hloramphenicol **a**cetyl**t**ransferase (CAT), and the firefly gene for luciferase. Two shared characteristics make these genes such useful experimental tools. One is that their products can be readily assayed. For example, *lacZ* encodes the enzyme β-galactosidase, which can be quantified because of its dose-dependent ability to generate a blue colour from the indicator dye X-Gal. It is also common to use S1 mapping, RNase protection or primer extension assays to quantify the abundance of transcripts derived from these reporters. The other useful feature of these reporter genes is that they are not normally present in the cells used to investigate eukaryotic gene expression. This means that effects are never masked by background expression from endogenous genes; if β-galactosidase mRNA or activity is detected in a eukaryotic cell, it must come from the reporter construct. By linking a promoter or enhancer fragment to a reporter gene, one can test the ability of that fragment to influence expression of the reporter. Alternatively, by linking a reporter to the known binding site of a transcription factor, one can test the effect of various manipulations on the activity of that factor.

novel elements that can respond to a range of different stimuli must have been crucial for the establishment of regulatory networks during the evolution of multicellular organisms.

DNA-binding factors that activate pol II transcription

Like promoters and enhancers, the transcription factors that bind to regulatory DNA sites of class II genes are generally found to be highly modular. These proteins can frequently be dissected into two or more separate domains, each of which is capable of carrying out an autonomous function. Thus, many of these proteins contain a DNA-binding domain that will function efficiently when it is removed from the remainder of the protein. They also tend to have one or more autonomous activator domains. A paradigm for this organization is provided by the yeast factor Gal4.

Gal4 is involved in controlling a group of genes which encode proteins that are required for the cell to utilize galactose as an energy source. These are referred to as *GAL* genes; they are virtually silent when yeast are grown in medium that contains no galactose, but are expressed very actively if galactose is available. This regulation is mediated by a promoter element called the

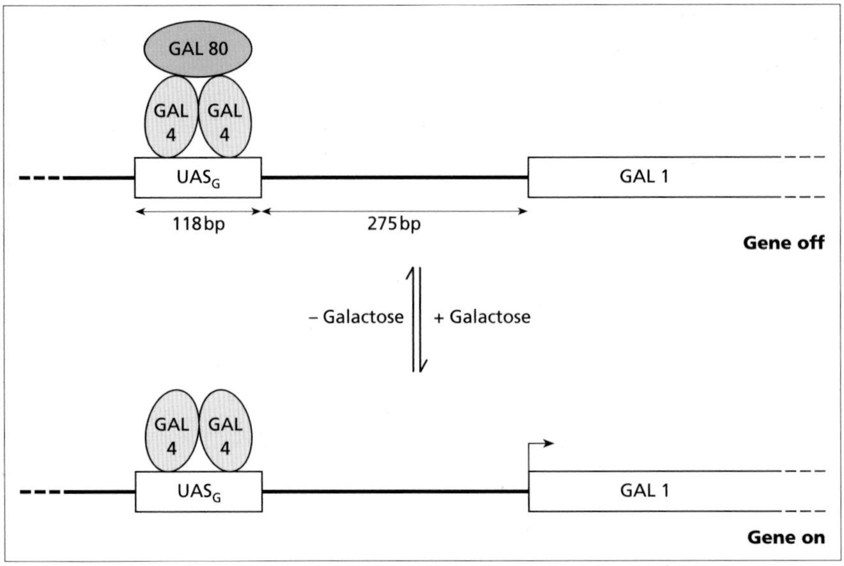

Fig. 5.3 Control of Gal4 activity. The *GAL1* gene in *S. cerevisiae* is regulated by a 118-bp UAS$_G$ located 275 bp upstream of its start site. This UAS$_G$ contains four 17 bp recognition sites, each of which is bound by a Gal4 dimer, and serves to activate *GAL1* transcription when galactose is present in the medium. In the absence of galactose the Gal80 repressor protein binds to Gal4 and blocks its ability to stimulate expression, even though it remains bound to the UAS$_G$.

galactose **u**pstream **a**ctivating **s**equence (UAS$_G$), which contains binding sites for several molecules of Gal4 protein. If the UAS$_G$ is deleted, *GAL* genes lose their galactose inducibility. Conversely, galactose responsiveness can be conferred on a heterologous gene simply by linking it to a UAS$_G$. The activator Gal4 binds the UAS$_G$ whether or not galactose is present. However, in the absence of galactose a repressor protein called Gal80 binds to Gal4 and prevents it from stimulating transcription (Fig. 5.3).

The modular nature of the Gal4 protein was revealed by deletion analysis and 'domain swap' experiments. Thus, deletion of all but the first 147 amino acids from the 881-residue full-length protein leaves a domain that is fully functional in DNA binding and dimerization, but is unable to activate transcription. The structure of this domain was described in Chapter 3. It can be further dissected to reveal an N-terminal fragment of 65 residues that can interact weakly with the UAS$_G$. Inclusion of residues 65–94 is necessary to allow efficient dimerization. When deprived of residues 1–147, Gal4 is no longer able to recognize DNA. However, if transferred to another DNA-binding domain, Gal4 residues 147–881 will confer the ability to stimulate gene expression. For example, a chimeric protein comprising Gal4 residues 147–881 linked to the DNA-binding domain of the bacterial protein LexA will activate transcription from promoters containing LexA-recognizition

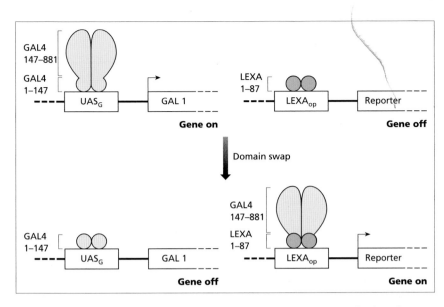

Fig. 5.4 A domain swap experiment. The isolated DNA-binding domains of Gal4 and LexA have little effect upon gene expression. However, either can stimulate transcription when linked to Gal4 residues 147–881, which include two activation regions. LexA$_{op}$ is the operator DNA sequence recognized by LexA.

Fig. 5.5 Functional regions of Gal4. The N-terminal 65 residues of Gal4 are capable of weak binding to the UAS$_G$. Inclusion of residues 65–94 allows efficient dimerization. Addition of residues 148–196 or 768–881 will confer the ability to activate transcription. Residues 851–881 are necessary for responsiveness to Gal80.

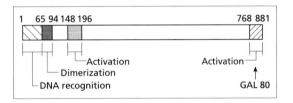

motifs (Fig. 5.4). Regions of Gal4 that can stimulate gene expression have been mapped to residues 148–196 and 768–881. Furthermore, the 30 residues at the extreme C-terminus are required for responsiveness to Gal80 (Fig. 5.5). It is clear from these experiments that Gal4 is composed of a series of discrete and autonomous functional modules.

Transcription activation domains

As just described, Gal4 contains two discrete activation domains. A striking feature that is shared by these two regions is that they are both markedly acidic; that is they contain substantially more acidic residues (Asp and Glu) than basic ones (Arg and Lys). At physiological pH, the C-terminal activation region (residues 768–881) has a net charge of –9 and the more N-terminal activation region (residues 148–196) has a net charge of –7. Either of these

VP16 = viral activator.

regions will stimulate expression when fused to a DNA-binding domain. Furthermore, an artificial protein composed of both these regions fused to the Gal4 DNA binding domain will activate transcription almost as efficiently as full-length Gal4, despite missing more than half of the intact protein. When domain swap experiments were used to map the activating regions of other transcription factors, many were also found to be enriched in acidic residues. Examples include the yeast activator GCN4, the human tumour suppressor p53 and the VP16 protein of herpes simplex virus. The C-terminal 80 residues of VP16 are highly acidic and constitute an especially powerful activation domain. When it is fused to the Gal4 DNA-binding domain (residues 1–147), the resultant chimera (called Gal4-VP16) will serve as a potent regulator of promoters that contain a UAS_G. This provides a particularly striking example of modularity, because a protein domain from yeast is able to function efficiently when combined with a polypeptide from a virus that infects people. The ability of these proteins to function when brought together across such a wide phylogenetic divide illustrates the fact that many features of the transcription apparatus are highly conserved. The organization of proteins in discrete structural modules provides a tremendous evolutionary advantage, because it allows natural selection to create novel functions by recombining pre-existing modules. Precisely the same consideration applies to the evolution of enhancers.

When randomly generated polypeptides from *E. coli* were fused to the DNA-binding domain of Gal4, about 1% of the resulting hybrids were found to stimulate expression in yeast of a UAS_G-dependent *lacZ* reporter gene. When fused to LexA, the same polypeptides activated promoters containing LexA-binding sites. In every case, the novel activating region was found to be acidic. However, there was no strict correlation between the degree of acidity and the strength of the activator. Some experiments suggest that the acidic residues need to be aligned along one face of the polypeptide in order to be most effective. Nevertheless, the frequency with which activation domains arise from random polypeptides implies that the sequence and structural requirements for this function are not very stringent. Indeed, acidic activating regions from different transcription factors generally show little or no similarity to each other, apart from their negative charge.

Many activation domains are not enriched in acidic residues. In a number of metazoan factors, a preponderance of glutamine residues is found. Examples include Sp1 and Oct-1, both of which contain two separate glutamine-rich activating regions (Fig. 5.6). The factors CTF and NTF-1 contain proline-rich and isoleucine-rich activation domains, respectively. Simple homopolymeric runs of glutamine or proline can stimulate transcription when fused to the DNA-binding domain of Gal4. Maximal activity is obtained with 10–30 glutamines or about 10 prolines. Polyproline and polyglutamine stretches are often encoded by multiple repetitions of the same codon in regions of DNA called triplet repeats or microsatellites. Triplet repeats can

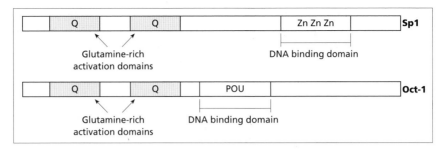

Fig. 5.6 Functional regions of Sp1 and Oct-1. Both factors contain a pair of glutamine-rich activation domains in the N-terminal half of the protein. The DNA-binding domain of Sp1 consists of three zinc fingers of the C_2H_2 class. A POU domain is responsible for DNA-binding by Oct-1.

evolve extremely rapidly and are sometimes highly unstable. This instability has been implicated in a number of genetic diseases, such as myotonic dystrophy.

The requirements for forming an activation domain are still not understood. In general, the positions of the negatively charged residues in acidic activation domains appear to be unimportant. When most of the negative charges were removed from Gal4 by truncations and missense mutations, its ability to activate was only moderately compromised. Furthermore, the addition of positively charged residues has little effect on Gal4 function. For both acidic and glutamine-rich domains, mutagenesis suggests that bulky hydrophobic residues interspersed between the predominant amino acid species may be of prime importance for the activation function. Because activating domains appear to operate by making protein–protein contacts with other components of the transcription apparatus, it is likely that these interactions are mainly hydrophobic, involving solvent exclusion and Van der Waals contacts. Specificity may then be determined by the periodicity of the cohesive elements. It is possible that acidic residues in activation domains establish long-range electrostatic interactions which increase the rate at which the target is located. In some but not all cases, studies of isolated activation domains by nuclear magnetic resonance (NMR) or circular dichroism have provided no evidence of secondary structure. It is thought that these domains may remain unstructured until they undergo an induced fit on binding to a specific target. Indeed, NMR analysis has shown that the minimal activation region of VP16 undergoes an induced transition from random coil to α-helix when it binds to one of its targets $hTAF_{II}31$. Such conformational restructuring may reduce the danger of non-productive interactions under inappropriate conditions and may also allow for greater flexibility. The combined use of multiple interactions may ensure that the overall specificity of association is high, even if individual interactions have low specificity. This may provide greater scope for regulation. It also provides a more flexible system, as the combinatorial use of factors would be severely restricted if all interactions were highly specific. It is important to note that several distinct types of acidic or glutamine-rich activators may exist, because not all domains of a particular class interact with the same target.

Interactions between activators and the basal transcription machinery

The primary function of activation domains is to interact with other components of the transcription apparatus. In many cases this involves binding directly to one or more of the basal factors. In other cases the interaction is an indirect one, in which the activation domain binds to a protein called a cofactor which, in turn, binds to the basal apparatus. Direct interactions with the basal machinery will be described first.

Many activators, of both viral and cellular origin, have been shown to bind directly to one or more components of TFIID. As described in the last chapter, TFIID is a multisubunit complex composed of TATA-binding protein (TBP) and at least 10 TAFs (TBP-associated factors). Separate regions of the acidic activation domain of VP16 bind specifically to TBP and to $hTAF_{II}31$ or its *Drosophila* homologue $dTAF_{II}40$. Another acidic activation domain, that of p53, can bind to TBP, $hTAF_{II}31/dTAF_{II}40$ and $hTAF_{II}80/dTAF_{II}60$. The glutamine-rich activation domains of Sp1, CREB and Bicoid bind to $hTAF_{II}130$ or its *Drosophila* homologue $dTAF_{II}110$. Yeast appear to lack a homologue of $hTAF_{II}130/dTAF_{II}110$, and it is striking that glutamine-rich regions will not stimulate transcription in yeast. Although a particular TAF is often targetted by several members of a particular class of activator, individual TAFs may also be recognized by distinct types of activator. For example, $dTAF_{II}60$ is bound by the acidic activation domain of p53, but also by the isoleucine-rich activation domain of the *Drosophila* factor NTF-1.

Apart from TFIID, several other components of the basal pol II transcription apparatus have been implicated as direct targets for activator proteins, including TFIIB, TFIIF and TFIIH. Indeed, the activation domain of VP16 has been shown to bind to TFIIB and TFIIH, as well as TBP and $dTAF_{II}40$. It is highly unlikely that a single VP16 molecule could interact with each of these targets simultaneously. However, if several molecules of VP16 were associated with a promoter, they might together be able to influence multiple components of the basal machinery.

A convenient strategy that has been commonly used to identify targets for activators is to generate immobilized beads that carry multiple copies of a covalently linked activation domain. Potential target proteins are then mixed with these beads in order to assay for binding. After a sufficient incubation period, unbound proteins are washed away whereas factors that interact with the activation domain remain associated with the beads. The potential target can be detected by autoradiography, if it has been radiolabelled, or by western blotting if an antibody is available. Specific retention by the beads suggests that

Box 5.2 Pull-down assay

The pull-down assay is used to test whether two proteins interact *in vitro*. The most common version involves producing one of the proteins of interest (X) as a recombinant fusion with bacterial **glutathione S-t**ransferase (GST). The assay then exploits the ability of GST to bind with very high affinity to commercially available agarose beads that are coated with glutathione. The GST-X fusion protein is immobilized on these beads and then mixed with the test protein (Y). After an appropriate incubation, the beads are pelleted by centrifugation and the supernatant is removed. The presence of Y is then determined by western blotting or autoradiography. If Y binds to GST-X, it will be found in the pellet with the beads; if it does not bind to X, it will be found in the supernatant. An important specificity control is to test in parallel whether Y associates with beads carrying GST alone.

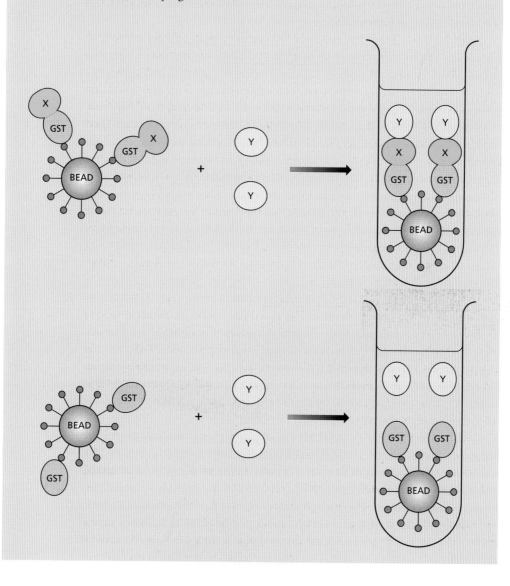

expts - fn
Activator ↓ targets

a factor is bound by that particular activation domain. When carried out on a small scale, this approach is often referred to as a pull-down assay (Box 5.1); it can also be scaled up and used as affinity chromatography, with the beads held in a column. Unfortunately, some of the interactions that have been revealed by this strategy are likely to be artefactual. A major problem is that the assays are carried out using concentrations of proteins that are orders of magnitude greater than would ever occur *in vivo*. Such high concentrations can force interactions that would not occur under physiological conditions. An additional problem is that the proteins have been removed from their natural context, and this may expose regions of polypeptide that would be buried in a functional transcription complex. It is therefore essential to employ additional tests in order to verify the physiological relevance of a potential target that has been implicated using protein-binding studies.

Problem

One way to confirm the relevance of a protein–protein interaction is to mutagenize the binding site and test what effect this has upon function. If a mutation that abolishes binding has no effect upon transcription, then the binding is unlikely to be important. For Gal4, mutations in the activation domain that diminish the interaction with TBP or TFIIB also diminish the ability to stimulate transcription. In the case of VP16, there is also generally a good correlation between binding and function. Thus, substitutions in the acidic activation domain that prevent it from stimulating transcription also prevent the interactions with TFIID, TFIIB and TFIIH. Conversely, mutant forms of TFIIB have been isolated that no longer bind to the acidic tail of VP16; these mutants will only support basal levels of transcription that can no longer respond to the presence of VP16. However, certain TBP mutations that prevent the binding of VP16 or p53 *in vitro* have no effect on activation by these factors *in vivo*.

An alternative approach that has been used to verify the functional significance of *in vitro* binding assays has involved reconstituting transcription in the presence or absence of the putative target. For example, pull-down assays showed that dTAF$_{II}$110 is bound by the glutamine-rich activation domain of the *Drosophila* factor Bicoid, whereas an adjacent alanine-rich region interacts with dTAF$_{II}$60. A partial TFIID complex composed of just TBP, dTAF$_{II}$250 and dTAF$_{II}$110 was found to support activation by the glutamine-rich domain, but not the alanine-rich domain, whereas a complex composed of TBP, dTAF$_{II}$250 and dTAF$_{II}$60 allows activation by the alanine-rich domain of Bicoid but not its glutamine-rich domain (Fig. 5.7). Thus, transcriptional stimulation only occurred when the appropriate TAF target was present. This confirmed the functional significance of the protein-binding (pull-down) assays. It also confirmed that individual TAFs can serve as receivers for signals provided by distinct activation domains.

Many studies carried out *in vitro* using extracts of human or *Drosophila* cells have found that transcriptional activation is sensitive to the presence of appropriate TAFs. Thus TBP can substitute for the TFIID complex in supporting basal transcription from the adenovirus major late promoter, but many studies have

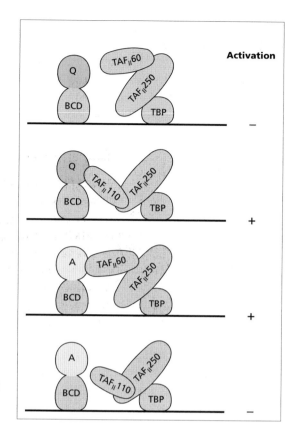

Fig. 5.7 An appropriate target is required for activation. Bicoid has two activation domains: a glutamine-rich activation domain (Q) that binds to dTAF$_{II}$110 and an alanine-rich activation domain (A) that binds to dTAF$_{II}$60. When alternative TFIID subcomplexes were tested, the appropriate TAF target was required in order to allow transcriptional stimulation by either activation domain.

failed to obtain activation in a purified system reconstituted without TAFs. These observations are consistent with the finding that TAFs are targetted by many activation domains. However, a surprise was obtained when genetic analysis was used to examine the role of TAFs in yeast cells. Various strategies were used to inactivate a range of yeast TAF$_{II}$s *in vivo*. Loss of individual TAF$_{II}$s caused the yeast to die, which clearly establishes their importance to the organism. However, many genes continued to be expressed at high levels until the cells died. For example, activation by GCN4 was uncompromised when the levels of various TAFs were reduced to 5% of wild type. It is conceivable that the remaining 5% is sufficient to allow transcription to continue as normal, but this seems highly unlikely. The abundance of individual TAF$_{II}$s in *S. cerevisiae* is thought to be similar to the number of pol II promoters (~6000 per cell). As the TAFs are not present in a large stoichiometric excess, one would expect a 20-fold decrease in their levels to affect most promoters at which they perform an essential function. It therefore appears that GCN4 can stimulate transcription in a TAF-independent fashion and similar results were obtained for a range of other yeast activators. This does not mean that TAFs are not involved in the normal process of activation in these cases; multiple mechanisms may permit activation, so that loss of any one pathway may be compensated by the other(s). Nevertheless, Gal4-dependent activation was re-duced by three- to four-fold and several promoters with weak TATA elements were substantially downregulated. These results are consistent with the idea that $TAF_{II}s$ mediate the response to particular activators and facilitate inter-actions with promoters that lack a strong TATA sequence. TAF-independent pathways of gene activation also exist in higher organisms, although they have sometimes been difficult to detect.

Cofactors

Although direct interactions between activators and the basal machinery are believed to make an important contribution to transcriptional stimulation, an additional class of factors is also considered to be fundamental to this process. These factors are referred to by a number of terms, including cofactors, adaptors and mediators; the variety of the labels reflects the heterogeneity of the group.

Squelching

One of the first indications of the existence of cofactors came from the phe-nomenon of squelching. Although adding an activator will normally increase transcription, when it is present in large excess it may actually suppress or squelch it. This is thought to occur because the excess activator molecules bind a target protein and sequester it in non-productive complexes, so that it can no longer perform its normal function (Fig. 5.8). Indeed, overexpression of a Gal4–VP16 fusion protein can be toxic in yeast. An excess of one factor can sometimes interfere with the stimulation of expression by another, suggesting

Mediator

competition between the two activators for a shared target. In yeast, squelch-ing can often be relieved by the addition of a multisubunit complex referred to as the mediator. The mediator is part of the pol II holoenzyme. It consists of about 20 polypeptides and associates with the C-terminal domain (CTD) of the largest subunit of pol II. Genes encoding some of these polypeptides were isolated as suppressors of a cold-sensitive phenotype that results from trunca-tion of the CTD. These genes were named *srb*, which stands for **s**uppressor of **R**NA polymerase **B**. Mutations in *srb2* or *srb5* have similar consequences to CTD truncations, consistent with the observed interaction between SRB pro-teins and the CTD. All of the *srb* genes are required for normal growth in yeast. As well as stimulating basal transcription, the mediator is involved in activa-tion by Gal4–VP16; if it is dissociated using antibodies against the CTD, the response to activators is compromised. SRB proteins are also present in higher eukaryotes, as is an involvement of the CTD in transcriptional activation. Many activators are unable to function in human cells expressing a pol II mutant with a deleted CTD.

CBP and p300 are a related pair of large adaptor proteins (>2400 residues) that play important roles as cofactors in metazoan organisms. For example, CBP mediates transcriptional activation by a factor called **c**AMP-**r**esponse **e**lement-**b**inding protein (CREB) in response to **c**yclic **AMP** (cAMP). Pro-moters that are regulated by cAMP often contain a DNA motif called the **c**AMP-**r**esponsive **e**lement or CRE. This sequence can be recognized by a family of

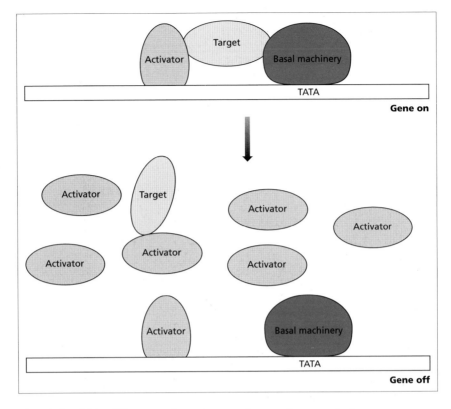

Fig. 5.8 Squelching. When an activator is present in excessive amounts, it can suppress transcription by sequestering a target protein in non-productive complexes.

at least 10 different factors with bZIP DNA-binding domains, one of which is CREB. CREB contains a glutamine-rich activation domain that functions constitutively by binding to hTAF$_{II}$130. It also contains an inducible activation domain that only becomes operative after it has been phosphorylated by the cAMP-dependent protein kinase A (PKA). Various extracellular stimuli cause adenylate cyclase at the plasma membrane to convert ATP into cAMP. Binding of cAMP to the regulatory subunits of PKA releases the catalytic subunits, which move to the nucleus and phosphorylate CREB on serine-133 within its inducible activation domain. Phosphorylated CREB is recognized by CBP, which can form a bridge to the pol II holoenzyme (Fig. 5.9). Thus, CREB can communicate with pol II, but only after it is phosphorylated at Ser-133 in response to cAMP. In addition to its bridging function, CBP also plays an important role in opening up chromatin to increase the accessibility of promoters. This activity will be described later. The importance of CREB phosphorylation *in vivo* has been illustrated using transgenic mice in which Ser-133 is substituted with alanine; because cAMP serves as an important mitogenic signal for anterior pituitary somatotrophic cells, mice expressing the non-phosphorylatable Ala-133 mutant form of CREB display a dwarf phenotype with atrophied pituitary glands.

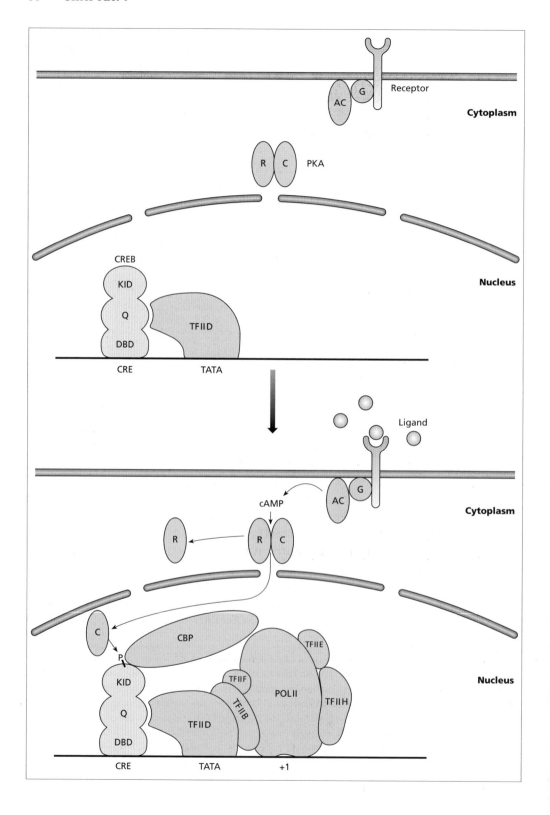

Apart from CREB, a number of other DNA-binding factors make use of CBP and p300 to stimulate transcription. For example, the nuclear hormone receptors interact with CBP/p300 through their ligand-binding domains, and this association is stimulated strongly by hormones. The AP-1 factors, which are composed of members of the Fos and Jun families, also utilize CBP and p300. In fact, competition for these cofactors appears to allow cross-talk between different groups of transcription factors and the signalling pathways that control them. Thus, nuclear hormone receptors can antagonize the effects of AP-1, but this phenomenon can be abrogated if CBP or p300 are over-expressed. Similarly, the ability of retinoic acid receptor (RAR) to suppress activation by AP-1 can be blocked using a synthetic ligand that prevents the RAR from binding to CBP/p300. Conversely, CREB which has been phosphory-lated at Ser-133 can suppress the induction of genes by retinoic acid. Different nuclear hormone receptors can also antagonize each other in a ligand-dependent manner. These instances, in which unrelated transcription factors compete for shared cofactors, provide strong evidence for the physiological importance of squelching. It may provide a mechanism for switching between different response pathways. Indeed, loss of one CBP allele appears to cause Rubinstein–Taybi syndrome, an autosomal dominant inheritable disease invol-ving mental retardation, unusually broad thumbs and facial abnormalities. One presumes that the level of CBP in these individuals is insufficient to allow the normal functioning of critical gene expression pathways.

Some cofactors are only found in particular tissues or developmental stages. An example is provided by OBF-1, a cofactor that interacts with the octamer-binding factors Oct-1 and Oct-2 in B-lymphocytes. The octamer motif (ATGCAAAT) is found in every promoter and most of the enhancers of the immunoglobulin genes, and is important for their B-cell-specific expres-sion. Whereas Oct-1 is present in most if not all cells, Oct-2 is only found in lymphocytes. When tested in HeLa cells, Oct-2 can activate immunoglobulin gene transcription, but Oct-1 cannot. However, in B cells the cofactor OBF-1 interacts with Oct-1 and allows it to stimulate the expression of immuno-globulin genes. Thus, at least two distinct mechanisms contribute to the tissue-specific production of immunoglobulins: the B-cell-specific activator Oct-2; and the B-cell-specific cofactor OBF-1 that allows the ubiquitous factor Oct-1 to function at immunoglobulin promoters and enhancers (Fig. 5.10).

Fig. 5.9 (*opposite*) Regulation of CREB in response to cAMP. When cAMP levels are low, the catalytic subunits (C) of protein kinase A (PKA) are inhibited by its regulatory subunits (R); CREB remains underphosphorylated and shows only basal activity, involving contact between its glutamine-rich domain and TFIID. Various extracellular ligands binding to cell surface receptors can activate adenylate cyclase (AC) via G proteins so that it synthesizes cAMP; cAMP causes release of the catalytic subunits of PKA, which translocate to the nucleus and phosphorylate Ser-133 within the kinase-inducible domain (KID) of CREB; when phosphorylated at this site, CREB binds CBP which activates transcription; CBP employs at least two distinct mechanisms for activation, one of which is the recruitment of pol II.

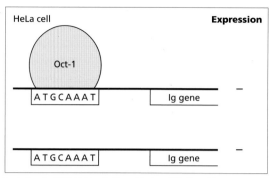

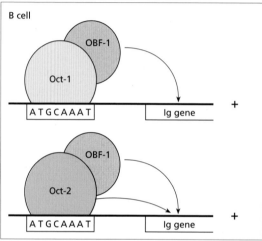

Fig. 5.10 Tissue-specific factors allow the expression of immunoglobulin (Ig) genes in B cells but not in HeLa cells. Oct-1 is present in HeLa cells, but is unable to stimulate Ig synthesis. B cells contain the cofactor OBF-1, which allows Oct-1 to function at Ig promoters and enhancers. B cells also produce Oct-2, which binds OBF-1 but does not require it to activate Ig gene transcription.

VP16 can be regarded as a cofactor that enables Oct-1 to regulate specific viral genes. When **h**erpes **s**implex **v**irus (HSV) infects a cell, five of its genes are transcribed very rapidly without the need for *de novo* protein synthesis. The promoters of these five 'immediate early' genes contain the sequence TAATGARAT (the R stands for purine), which allows them to be activated by VP16. VP16 enters the cell as part of the virion particle and does not bind independently to DNA. Instead, Oct-1 binds to the TAATGARAT motif, because the homeodomain portion of its POU DNA-binding domain recognizes the TAAT sequence as being similar to the AAAT of a normal octamer sequence (ATGCAAAT). However, the presence of the flanking GARAT element causes the POU domain to adopt a novel conformation. Only when it is in this conformation can Oct-1 recruit VP16. In this way, VP16 can discriminate against Oct-1 bound to cellular promoters with octamer motifs and specifically activate the immediate early viral genes where Oct-1 is present in a distinct conformation. An additional **h**ost-**c**ell **f**actor (HCF) is also assembled onto the TAATGARAT sequences and is necessary for recruitment of VP16 (Fig. 5.11). Thus, the virus provides its own cofactor in order to subvert the host's transcriptional apparatus. This greatly improves the efficiency with which HSV can infect

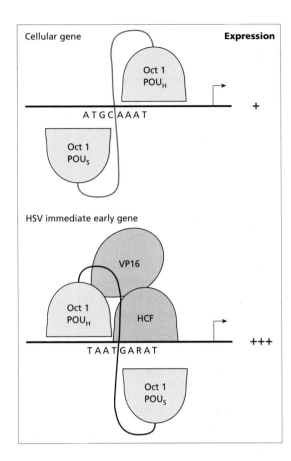

Fig. 5.11 The promoters of HSV immediate early genes contain a sequence TAATGARAT that is recognized by Oct-1 which contains POU_H and POU_S DNA binding domains; when bound to such sequences Oct-1 adopts a conformation which, in concert with host-cell factor (HCF), allows it to recruit VP16 with its powerful activation domain. VP16 does not recognize Oct-1 when it is bound to the octamer sequence ATGCAAAT found in cellular promoters or enhancers.

cells. However, the primary function of VP16 is as a structural component of the virion, rather than a custom-made cofactor. Even though VP16 from HSV has one of the most powerful activation domains identified, it would appear that this has evolved as a subsidiary function for a structural protein. This provides a fine example of the adaptive efficiency of viruses.

Mechanisms of activation

The previous section has described how activation domains make contact with other components of the transcription apparatus. These interactions can stimulate gene expression in a variety of ways, as will now be described (Fig. 5.12).

Removal of repressors

A major reason why basal pol II transcription does not occur *in vivo* is that the basal factors compete very poorly with histones and other proteins that fold DNA up into chromatin. Because of its great size, a eukaryotic genome must be tightly folded in order to fit into a nucleus. On their own, the basal pol II

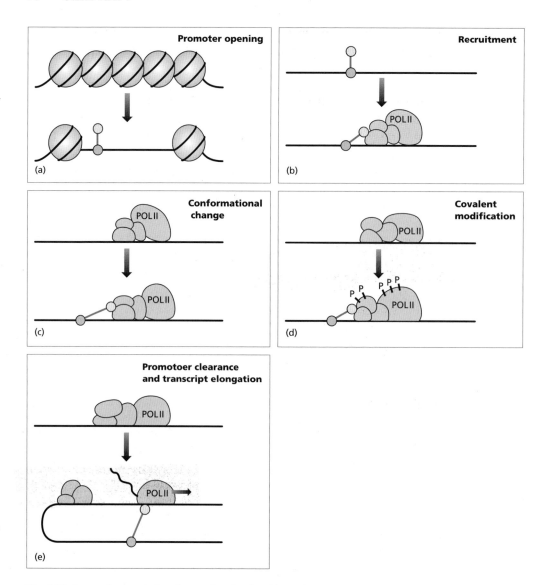

Fig. 5.12 Stages of transcription that can be stimulated by activators.

factors have great difficulty in gaining access to promoter DNA once it has been assembled into chromatin. This problem can be overcome with the assistance of various activating proteins. In many cases, the activators can find their recognition sequences even when they are presented as chromatin. There is also an opportunity for transcription factors to gain access to promoter or enhancer sequences shortly after the DNA has been replicated, because newly synthesized DNA is free from chromatin. Because TFIID binding to TATA sequences is slow and inefficient, it struggles to take advantage of the window of opportunity before chromatin reforms. However, many activators bind DNA rapidly and can occupy their recognition motifs before they become

obscured by chromatin. Once in position, such activators can remodel the surrounding chromatin so as to allow access of the basal factors.

As has already been described, CREB is able to recruit the cofactor CBP
once it has been phosphorylated in response to cAMP. In addition to its bridging role between CREB and the pol II holoenzyme, CBP performs an important function in 'opening-up' regions of chromatin. It is able to do this because
it has intrinsic histone acetyltransferase activity, which allows it to acetylate
the N-terminal tails of histones. These tails contain many lysine residues that
interact tightly with the negatively charged phosphodiester backbone of
DNA. Acetylation neutralizes the positive charge on the lysines, allowing the
DNA to unfold and become more accessible to transcription factors. As well as
having its own acetyltransferase activity, CBP recruits a second acetyltransferase
called P/CAF (**P**300/**C**BP-**a**ssociated **f**actor). CBP and P/CAF have subtly
different substrate specificity and together are thought to provide an effective
system for increasing the accessibility of chromatin domains. This is likely to
be an important prelude to transcription complex assembly at the core promoter region. Once the promoter is open, CBP assists in recruiting the pol II
holoenzyme (Fig. 5.13).

Because chromatin is believed to have a very substantial effect upon gene
expression *in vivo*, it will be described in detail in Chapter 8.

Recruitment of the basal factors and pol II

The best-characterized mechanism that is used by activators to stimulate transcription is the recruitment of basal factors. As already described, many activation domains make direct contact with components of the basal apparatus. In
other cases the interaction is indirect, involving intermediary cofactors. It is
believed that these interactions can serve to facilitate preinitiation complex
assembly. An incisive series of experiments in yeast have demonstrated the
importance of recruitment as a mechanism for activation. For example, the
DNA-binding domain of LexA on its own has no effect on transcription from
a promoter containing a LexA recognition sequence. If the DNA-binding
domain is linked to an acidic activation domain, expression is stimulated.
However, high levels of expression can also be achieved in this system if the
DNA-binding domain is coupled covalently to TBP (Fig. 5.14). In this case,
LexA brings TBP directly to the promoter; this concentrates TBP in the vicinity
of its binding site and thereby stimulates TATA box occupancy. Because DNA
recognition by LexA is much more efficient than TATA binding by TBP, recruitment of the latter is improved considerably. The LexA–TBP chimera gives
levels of expression that are comparable to those obtained with strong activation
domains, such as that of Gal4. Thus, efficient recruitment of a basal factor can
bypass the need for an activation domain. The result also shows that bringing
TBP to the promoter can be a rate-limiting step for transcription *in vivo*.

Many of the other basal factors are also contacted by specific activation
domains and this is likely to facilitate recruitment. Whether or not this will

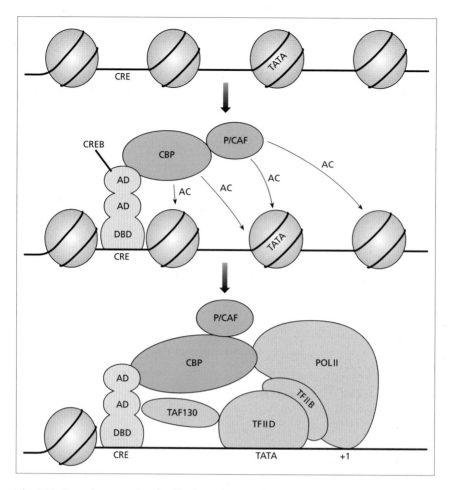

Fig. 5.13 Several steps are involved in the activation of transcription via a cAMP-responsive element (CRE). CREB can often bind to CRE sequences in the presence of nucleosomes, even though the TATA box is inaccessible. When phosphorylated by protein kinase A, CREB recruits the cofactor CBP, which can activate transcription by at least two distinct mechanisms. CBP has intrinsic histone acetyltransferase activity and also binds another acetyltransferase called P/CAF. By acetylating histones, CBP and P/CAF are thought to increase the accessibility of DNA in chromatin. CBP also performs a bridging role which helps recruit the pol II holoenzyme to open regions of chromatin. In addition, CREB itself has a second activation domain (AD) that makes contacts with TFIID via $TAF_{II}130$.

stimulate expression may depend upon circumstances. For example, if TATA box binding by TFIID is the rate-limiting step on a particular promoter, accelerating the recruitment of TFIIB might have little effect on the rate of assembly of the preinitiation complex. However, different steps in the assembly pathway are likely to be limiting under different conditions, probably in a promoter-dependent fashion. Because DNA binding by TFIID is the initial stage in assembling the basal machinery on a TATA-containing promoter, and as TATA binding is slow and inefficient, it seems probable that this step will be

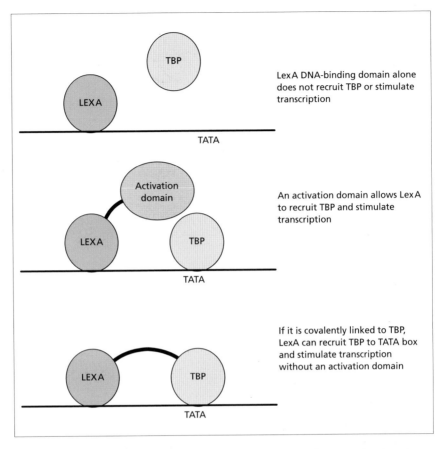

Fig. 5.14 Activated transcription can be achieved in the absence of an activation domain if TBP is coupled directly to an efficient DNA-binding domain. The DNA-binding domain alone of LexA does not affect transcription from a promoter containing a LexA recognition motif. If this LexA domain is coupled to the activation domain of Gal4, expression increases considerably. A similar effect can be obtained if the LexA DNA-binding domain is bound directly to TBP. The LexA–TBP hybrid allows efficient recruitment of TBP to the promoter, which results in high levels of expression.

limiting for the expression of many genes that were previously silent; this can explain why TFIID is targetted by so many activation domains. However, once TFIID is in place it remains bound through multiple rounds of transcription. By contrast, most of the other basal factors, such as TFIIB, dissociate from the promoter after each initiation event. Recruitment of TFIIB will therefore be necessary prior to reinitiation and may well become rate limiting. The ability of an activator to recruit TFIID may then be of no significance for the second round of transcription. Of course, an activator that can recruit the pol II holoenzyme will have a substantial advantage, because it will be able to bring in simultaneously many components of the basal machinery. Indeed, artificial interactions between the DNA-binding domains of LexA or Gal4 and components of the holoenzyme, such as SRB proteins, can activate transcription extremely efficiently.

A corollary of these 'activation domain bypass' experiments is that tethering basal factors to a promoter can overcome the repressive effects of chromatin. This has been demonstrated directly in the case of the yeast *PHO5* gene, which is regulated by a factor called Pho4. An array of histones is positioned at the *PHO5* promoter, and these are modified or removed when the gene is activated. When the activation domain of Pho4 is replaced by a component of the holoenzyme, the *PHO5* gene is expressed and chromatin remodelled as usual. In this case, chromatin reorganization is a direct consequence of recruiting the basal machinery. However, in other situations additional accessory factors, such as histone acetylases, may be necessary to open up regions of chromatin.

Conformational changes

Activation can also occur at a stage following recruitment of the basal factors and involve conformational changes in the assembled preinitiation complex. For example, the presence of activators can alter the footprint of TFIID dramatically, which suggests that rearrangements may occur. Gal4–VP16 has been shown to induce a conformational change in TFIIB, as revealed by changes in its susceptibility to proteolytic degradation. In the absence of activator, TFIIB is thought to adopt a configuration in which the docking sites for TFIIF and pol II are obscured. Gal4–VP16 appears to provoke a rearrangement in TFIIB which exposes the surfaces that are recognized by TFIIF and pol II, thereby facilitating the next step in preinitiation complex assembly (Fig. 5.15). Thus, the activation domain of VP16 has two complementary effects on TFIIB: it stimulates its recruitment and it reorganizes it into an active configuration.

Covalent modifications

Several components of the basal transcription apparatus have been shown to undergo phosphorylation and/or acetylation. It is likely that some of these covalent modifications may serve to regulate transcription. The Tat protein of human immunodeficiency virus (HIV) provides an example of an activator that stimulates transcription through phosphorylation of pol II. Tat is a highly unusual activator, because instead of binding to DNA it interacts specifically with a sequence at the 5' end of the nascent viral transcript called the transactivation response RNA element. Tat recruits a cellular kinase complex which phosphorylates the CTD of pol II. This kinase, which has been named TAK (Tat-associated kinase), is normally involved in increasing the rate of transcript elongation from cellular genes. Tat also binds to TFIIH and stimulates its ability to phosphorylate the CTD. Because hyperphosphorylation of the CTD is associated with promoter clearance and productive elongation, Tat can have a strongly stimulatory effect on the rate of transcription of the HIV genome.

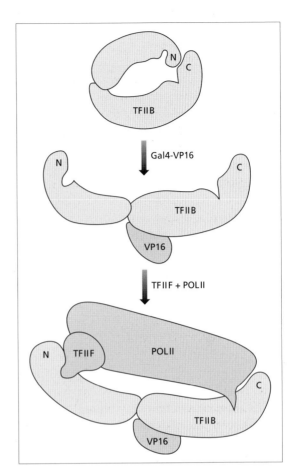

Fig. 5.15 Gal4–VP16 is thought to induce a conformational change in TFIIB which facilitates the binding of TFIIF and pol II. In the absence of the activator TFIIB is thought to be folded so that the N-terminal docking site for TFIIF and the C-terminal docking site for pol II are obscured. Binding of Gal4–VP16 triggers a rearrangement that allows access of TFIIF and pol II.

Stimulation of promoter clearance and elongation

The HIV Tat protein provides an example of an activator that primarily operates after transcriptional initiation. Both promoter clearance and transcript elongation can be regulated by specific activators. By stimulating promoter clearance—the step at which pol II is released from the initiation complex—an activator will accelerate not only the rate of RNA synthesis but also the rate at which another polymerase molecule can be recruited to the start site. For many genes transcribed by pol II, transcript elongation can be rate limiting and it is therefore targetted by various regulators. Control at the level of clearance or elongation offers the opportunity of a rapid response, because the initiation complex has already assembled.

Expression of many class II genes can be limited by a phenomenon called promoter-proximal pausing, in which the elongation complex stalls within 30–50 bp of the start site. The mammalian oncogene *c-myc* and the *Drosophila hsp70* gene provide well-characterized examples of this. In the latter case, the duration of promoter-proximal pausing is greatly diminished by an activator called the heat-shock factor (HSF). VP16 has also been shown to stimulate transcript elongation. However, by no means all activators are able to function at this level. Proteins that can accelerate elongation, such as VP16, HSF and Tat, probably do so by recruiting or stimulating elongation factors that improve the processivity of pol II. Evidence for this possibility is provided by the observation that an excess of the VP16 activation domain can squelch elongation. Hyperphosphorylation of the CTD has been implicated strongly as a modification that converts non-processive pol II into a processive state. The ability of Tat to increase CTD phosphorylation has already been described. Pol II molecules that have stalled near the start of non-activated *hsp* genes have hypophosphorylated CTDs, but these become hyperphosphorylated when expression is activated by heat shock. The ability of VP16 to stimulate elongation may be due, at least in part, to its affinity for TFIIF, because the latter can increase the overall rate of RNA polymerization.

Action at a distance

The fact that enhancer sequences can be located several kilobases from the promoter region raised the question as to how proteins bound to such remote sites are able to influence the behaviour of the basal transcription apparatus. Several types of mechanism could be envisaged for this. For example, the effects of enhancer-binding proteins might be transmitted to the promoter through changes in DNA structure, such as supercoiling. Alternatively, a factor could bind at a distant DNA site and then track along the DNA until it finds a promoter. Clear evidence of such a tracking mechanism has been found in the case of the bacteriophage T4 late genes. However, activation at a distance is thought most commonly to be achieved by looping out the intervening DNA, thereby allowing physical contact between proteins bound at spatially separate sites. Evidence in support of this mechanism came from the demonstration that an enhancer can stimulate transcription at a promoter on a separate DNA molecule if the two sites are closely associated by protein-mediated tethering or DNA catenation. In addition, electron microscopy has allowed the direct visualization of protein–protein interactions between various factors. For example, Sp1 molecules bound to sites ~1.8 kb apart can be seen to associate with each other, with the intervening DNA present as a large loop (Fig. 5.16). This capacity of Sp1 to interact across large distances requires a functional activation domain and correlates with the ability of spatially separate sites to stimulate expression synergistically *in vivo*.

The process of looping may be facilitated by curvature in the DNA between the two sites. The bend must be aligned so that proteins are moved together rather than apart. Such curvature may be an intrinsic property of a particular DNA sequence. Alternatively, it may be imposed by specific proteins that bend DNA. One very important way in which this can occur is through folding of the intervening DNA into chromatin, which can bring distal sites into proximity. This phenomenon will be described in more detail in Chapter 8. Certain transcription factors also function primarily by bringing together spatially

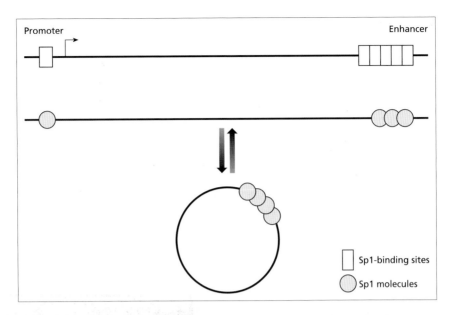

Fig. 5.16 Sp1 molecules bound at sites 1.8 kb apart can associate by looping out the intervening DNA. Electron microscopy was used to visualize nucleoprotein complexes that form on a DNA template containing a promoter with a single Sp1 site near one end and a cluster of six Sp1 sites at the other end (top). On some template molecules, Sp1 bound both ends but these remained separated. However, the ends of many DNA molecules were brought together by association of Sp1, with the intervening DNA appearing as a loop.

separate sites. The unifying feature of such 'architectural' transcription factors is that they remodel DNA regions by imposing severe bends at their sites of binding. A good example is provided by the lymphoid enhancer-binding factor 1 (LEF-1), which is found in pre-B and T lymphocytes. LEF-1 has a high mobility group (HMG) DNA-binding domain which uses a non-aromatic, hydrophobic amino acid to intercalate into a base step, thereby creating a ~130° bend in the DNA. Within the context of the T-cell receptor α gene enhancer, the LEF-1-induced bend allows interaction of at least two additional transcription factors that would otherwise be apart (Fig. 5.17). Mutations that alter the relative positions of the three factor recognition motifs inactivate the enhancer. The effect of an enhancer is sometimes restricted to a single promoter, even if several other promoters are located nearby; this restriction may be imposed by the precise stereospecific requirements for long-range interactions.

Synergism

It is frequently observed that multiple activators working together on a particular gene can have a synergistic effect on the expression of that gene. This means that two or more activator molecules produce a level of transcription that is greater than the sum of the levels achieved with each molecule acting alone. For example, in *Drosophila* the homeodomain protein Bicoid and the zinc

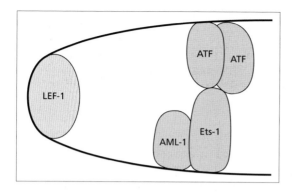

Fig. 5.17 LEF-1 binds to the enhancer of the T-cell receptor α gene and bends the DNA by ~130°. This allows cooperative interactions between the transcription factors ATF and Ets-1, which would otherwise be separated from each other.

finger protein Hunchback interact synergistically to activate the *hunchback* (*hb*) gene. When a *hb* promoter fragment was tested *in vitro*, either Bicoid or Hunchback alone increased transcription by ~six-fold, whereas together they produced a ~65-fold stimulation. Although some factors are highly selective about their choice of partners, others will synergise with a wide range of partners. For example, Gal4 can even synergise with mammalian factors such as the glucocorticoid receptor. Synergism is a very important phenomenon, because it provides a means of coordinating responses from many different signals. It can also amplify small changes in activator concentrations so as to give a much greater change in gene expression.

The effect of synergism can be achieved in a number of different ways. One way is for a pair of factors to show cooperativity in their binding to DNA. For example, in yeast cells in which the Gal4 concentration is low, a Gal4 dimer bound to a high-affinity DNA site can help a second Gal4 dimer bind to an adjacent low-affinity site. However, many factors can still work synergistically under conditions in which their DNA-binding sites are saturated. This phenomenon may be explained by the fact that a given activator can often interact with multiple components of the basal transcription apparatus, directly and/or via cofactors. For example, VP16 has been shown to bind to TBP, $TAF_{II}40$, TFIIB, TFIIF, TFIIH and the cofactor PC4. It is highly unlikely that a single molecule of VP16 could make all these interactions simultaneously, but if six VP16 molecules were arranged on a promoter they could each interact with one of the target factors (Fig. 5.18). The existence of preassembled holoenzyme complexes containing multiple components of the basal transcription apparatus will provide a contiguous target surface allowing the simultaneous action of several activator proteins.

Evidence that synergism can be achieved by the interaction of several activation domains with distinct targets came from experiments with various TFIID subcomplexes. As was mentioned previously, the glutamine-rich activation domain of Bicoid (BCD-Q) binds to $dTAF_{II}110$, whereas the alanine-rich domain (BCD-A) binds to $dTAF_{II}60$. No synergism is observed when two copies of BCD-Q or two copies of BCD-A are present at a promoter. However, one copy of BCD-Q combined with one copy of BCD-A together give syner-

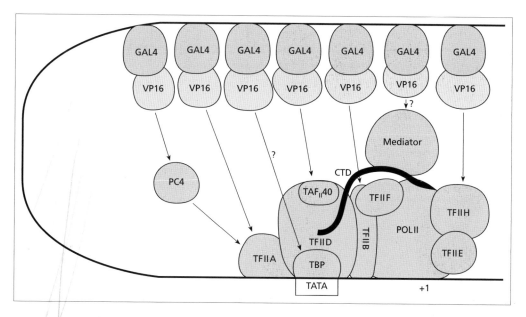

Fig. 5.18 Multiple molecules of an activator might function synergistically by contacting several components of the basal machinery simultaneously. For example, VP16 can bind to TBP, $TAF_{II}40$, TFIIB, TFIIF, TFIIH and the cofactor PC4; six molecules of Gal4-VP16 bound to a cluster of UAS_G elements might therefore stimulate transcription synergistically by contacting all of these targets.

gistic stimulation, but only if both $dTAF_{II}110$ and $dTAF_{II}60$ are present. These observations support the idea that synergism can be achieved by separate activation domains interacting with distinct targets; it is not sufficient to have several activators competing for the same target. In these experiments, the level of activation correlated with the degree of TATA box occupancy, which suggests that TFIID recruitment is responsible for the effects.

Kinetic
Synergism

If two separate slow steps (a and b) are necessary for transcription, then a phenomenon called kinetic synergism may occur. If activator X accelerates step a and activator Y accelerates step b, then little change in the rate of initiation will be obtained in the presence of either activator alone, because a slow step will remain. However, X and Y together will accelerate both slow steps leading to a synergistic increase in transcription. Recruitment of TFIID and the conformational rearrangement of TFIIB might represent two such slow steps on many pol II promoters.

Further reading

Reviews

Bentley, D. L. (1995) Regulation of transcriptional elongation by RNA polymerase II. *Curr. Opin. Genet. Dev.* **5**, 210–216.

Berk, A. J. (1999) Activation of RNA polymerase II transcription. *Curr. Opin. Cell Biol.* **11**: 330–335.

Bjorklund, S. & Kim, Y.-J. (1996) Mediator of transcriptional regulation. *Trends Biochem. Sci.* **21**: 335–337.

Buratowski, S. (1995) Mechanisms of gene activation. *Science* **270**: 1773–1774.

Carey, M. (1998) The enhanceosome and transcriptional synergy. *Cell* **92**: 5–8.

Dynan, W. S. (1989) Modularity in promoters and enhancers. *Cell* **58**: 1–4.

Firulli, A. B. & Olson, E. N. (1997) Modular regulation of muscle gene transcription: a mechanism for muscle cell diversity. *Trends Genet.* **13**: 364–369.

Hahn, S. (1993) Structure(?) and function of acidic transcription activators. *Cell* **72**: 481–483.

Herschlag, D. & Johnson, F. B. (1993) Synergism in transcriptional activation: a kinetic view. *Genes Dev.* **7**: 173–179.

Jones, K. A. (1997) Taking a new TAK on Tat transactivation. *Genes Dev.* **11**: 2593–2599.

Mannervik, M., Nibu, Y., Zhang, H. & Levine, M. (1999) Transcriptional coregulators in development. *Science* **284**: 606–609.

Parvin, J. D. & Young, R. A. (1998) Regulatory targets in the RNA polymerase II holoenzyme. *Curr. Opin. Genet. Dev.* **8**: 565–570.

Ptashne, M. (1992) *A Genetic Switch*, 2nd edn. Blackwell Scientific Publications, Cambridge, MA.

Ptashne, M. & Gann, A. (1997) Transcriptional activation by recruitment. *Nature* **386**: 569–577.

Ranish, J. A. & Hahn, S. (1996) Transcription: basal factors and activation. *Curr. Opin. Genet. Dev.* **6**: 151–158.

Rippe, K., von Hippel, P. H., & Langowski, J. (1995) Action at a distance: DNA-looping and initiation of transcription. *Trends Biochem. Sci.* **20**: 500–506.

Serfling, E. Jasin, M. & Schaffner, W. (1985) Enhancers and eukaryotic gene transcription. *Trends Genet.* **1**: 224–230.

Shikama, N., Lyon, J. & La Thangue, N. B. (1997) The p300/CBP family: integrating signals with transcription factors and chromatin. *Trends Cell Biol.* **7**: 230–236.

Triezenberg, S. J. (1995) Structure and function of transcriptional activation domains. *Curr. Opin. Genet. Dev.* **5**: 190–196.

Werner, M. H. & Burley, S. K. (1997) Architectural transcription factors: proteins that remodel DNA. *Cell* **88**: 733–736.

Yankulov, K. & Bentley, D. (1998) Tat cofactors and transcriptional elongation. *Curr. Biol.* **8**: R447–R449.

Zawel, L. & Reinberg, D. (1995) Common themes in assembly and function of eukaryotic transcription factors. *Annu. Rev. Biochem.* **64**: 533–561.

Selected papers

Enhancers

Ondek, B., Gloss, L. & Herr, W. (1988) The SV40 enhancer contains two distinct levels of organization. *Nature* **333**: 40–45.

Ondek, B., Shepard, A. & Herr, W. (1987) Discrete elements within the SV40 enhancer region display different cell-specific enhancer activities. *EMBO J.* **6**: 1017–1025.

Modularity in transcription factors

Brent, R. & Ptashne, M. (1985) A eukaryotic transcriptional activator bearing the DNA specificity of a prokaryotic repressor. *Cell* **43**: 729–736.

Acidic activation domains

Ma, J. & Ptashne, M. (1987) A new class of yeast transcriptional activators. *Cell* **51**: 113–119.

Targets for activation domains

Roberts, S. G. E., Ha, I., Maldonado, E., Reinberg, D. & Green, M. R. (1993) Interaction between an acidic activator and transcription factor TFIIB is required for transcriptional activation. *Nature* **363**: 741–744.

Wu, Y., Reece, R. J. & Ptashne, M. (1996) Quantification of putative activator–target affinities predicts transcriptional activating potentials. *EMBO J.* **15**: 3951–3963.

Involvement of TAFs in transcriptional activation

Moqtaderi Z., Bai, Y., Poon, D., Weil, P. A. & Struhl, K. (1996) TBP-associated factors are not generally required for transcriptional activation in yeast. *Nature* **383**: 188–191.

Walker, S. S., Reese, J. C., Apone, L. M. & Green, M. R. (1996) Transcription activation in cells lacking TAF$_{II}$s. *Nature* **383**: 185–188.

Mechanisms of activation

Chatterjee, S. & Struhl, K. (1995) Connecting a promoter-bound protein to TBP bypasses the need for a transcriptional activation domain. *Nature* **374**: 820–822.

Choy, B. & Green, M. R. (1993) Eukaryotic activators function during multiple steps of preinitiation complex assembly. *Nature* **366**: 531–536.

Klages, N. & Strubin, M. (1995) Stimulation of RNA polymerase II transcription initiation by recruitment of TBP *in vivo*. *Nature* **374**: 822–823.

Li, X.-Y., Virbasius, A., Zhu, X. & Green, M. R. (1999) Enhancement of TBP binding by activators and general transcription factors. *Nature* **399**: 605–609.

Roberts, S. G. E. & Green, M. R. (1994) Activator-induced conformational change in general transcription factor TFIIB. *Nature* **371**: 717–720.

Yankulov, K., Blau, J., Purton, T., Roberts, S. & Bentley, D. (1994) Transcriptional elongation by RNA polymerase II is stimulated by transactivators. *Cell* **77**: 749–759.

Action at a distance

Dunaway, M. & Droge, P. (1989) Transactivation of the *Xenopus* rRNA gene promoter by its enhancer. *Nature* **341**: 657–659.

Muller, H.-P., Sogo, J. M. & Schaffner, W. (1989) An enhancer stimulates transcription in *trans* when attached to the promoter by a protein bridge. *Cell* **58**: 767–777.

Su, W., Jackson, S. P., Tjian, R. & Echols, H. (1991) DNA looping between sites for transcriptional activation: self-association of DNA-bound Sp1. *Genes Dev.* **5**: 820–826.

Theveny, B., Bailly, A., Rauch, C., Rauch, M., Delain, E. & Milgrom, E. (1987) Association of DNA-bound progesterone receptors. *Nature* **329**: 79–81.

Synergism

Carey, M., Lin, Y.-S., Green, M. R., & Ptashne, M. (1990) A mechanism for synergistic activation of a mammalian gene by GAL4 derivatives. *Nature* **345**: 361.

Kim, T. K. & Maniatis, T. (1997) The mechanism of transcriptional synergy of an *in vitro* assembled interferon-β enhanceosome. *Mol. Cell* **1**: 119–129.

Poellinger, L., Yoza, B. K., & Roeder, R. G. (1989) Functional cooperativity between protein molecules bound at two distinct sequence elements of the immunoglobulin heavy-chain promoter. *Nature* **337**: 573.

Sauer, F., Hansen, S. K. & Tjian, R. (1995) DNA template and activator–coactivator requirements for transcriptional synergism by *Drosophila* Bicoid. *Science* **270**: 1825–1828.

Chapter 6: Transcription by RNA Polymerase I

Although a typical eukaryotic cell may contain over 10 000 transcripts, pol I synthesizes only a single RNA species, the large **r**ibosomal **RNA** (rRNA) precursor molecule. The fact that pol I is only required to synthesize rRNA has been demonstrated genetically in yeast, where it is possible to maintain viability after inactivating pol I if the rRNA coding region is linked artificially to a pol II promoter. The large rRNA precursor molecule is processed into the mature 5.8S, 18S and 28S rRNA components of ribosomes. Although it synthesizes only one essential product, pol I may be responsible for as much as 70% of all nuclear transcription and in an actively growing cell rRNA may constitute 80% of the total steady-state RNA. This astonishing level of production is necessary in order to make the several million new ribosomes that are needed per generation to maintain protein synthetic capacity. Because pol I is such a major determinant of biosynthetic capacity, its activity is linked very tightly to the rate of cellular growth. In hyperproliferative diseases such as cancer, pol I transcription may run wild.

To facilitate production, the pre-rRNA is encoded by multiple gene copies. In different eukaryotes, the number ranges between 100 and 5000 rRNA genes per haploid genome; mammals have ~150–200 copies. In most species, this **r**ibosomal **DNA** (rDNA) is organized in tandem head-to-tail arrays, with each ~7–14-kb coding region separated by a non-transcribed 'spacer' of ~2–30 kb. However, certain unicellular eukaryotes have extrachromosomal copies of the rDNA. Furthermore, rDNA is amplified specifically in *Xenopus* oocytes, in order to allow these frogs to synthesize and stockpile vast excesses of rRNA which are utilized subsequently during early development. A very high density of polymerase is found at the active rRNA genes, with approximately one pol I molecule every 100 bp, and these sustain a rapid elongation rate of ~30 nucleotides per second. The nascent transcript is processed and assembled into ribosomes while it is being synthesized. This feverish activity can be visualized in electron micrographic sections called 'Miller spreads', where gradients of densely packed nascent ribonucleoprotein are seen to emerge from the rDNA, forming patterns reminiscent of Christmas trees (Fig. 6.1). Indeed, these ribosome factories can even be seen under a light microscope and are referred to as nucleoli. Nucleoli constitute the most obvious landmarks within the nucleus and are generally thought of as discrete organelles. However, they are not surrounded by a membrane and are dynamic structures that form when pol I is active and disperse again when transcription stops (e.g. during mitosis or when cells are treated with actinomycin D). Their dynamic nature is demonstrated by the fact that transferring rDNA to novel chromosomal locations

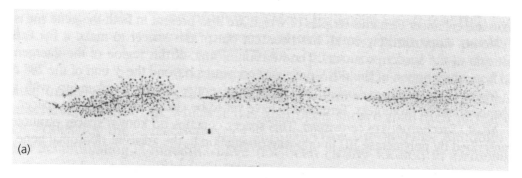

(a)

Fig. 6.1 Electron micrograph of *Xenopus* rDNA being transcribed. Each 'Christmas tree' corresponds to an rRNA gene, whereas the intervening spacer appears devoid of activity. Ribonucleoprotein fibrils emanate from the rRNA genes and a pol I molecule appears as an electron-dense granule at the base of each. Reprinted from *Trends in Genetics*, **6**, rRNA synthesis in nucleotides, pp. 390–395, copyright, with permission from Elsevier Science.

can result in the assembly of ectopic nucleoli. In very old cells, the nucleolus fragments into several separate nucleolar bodies and it has been suggested that the degeneration of nucleoli may be an important aspect of cellular ageing. The rDNA may be especially vulnerable to cumulative damage because it is organized into tandem repeats that are potentially unstable. Abnormally enlarged nucleoli are classic histopathological features of malignant cancer cells.

As well as the rRNA, ribosomes contain approximately 85 proteins. These are synthesized in the cytoplasm and transported to the nucleolus for assembly. They also contain a single molecule of the ~120-nucleotide 5S rRNA, which is made in the nucleoplasm by pol III. It is a perplexing anomaly that the 5S rRNA is synthesized by a different polymerase from the rest of the rRNA. 5S, 5.8S, 18S and 28S rRNA are all required in equal stoichiometry, each being present in one copy per ribosome. To synthesize them as a single precursor would provide an efficient and economical way of ensuring that they are always produced in the correct ratio. This would spare the cell from the metabolic expense of producing surplus transcripts, which should be beneficial considering the huge amounts of rRNA that are manufactured. Nevertheless, despite these obvious advantages, 5S rRNA is made in excess by a different polymerase.

The organization of rRNA genes

The important promoter sequences of rRNA genes from several species were determined initially by *in vitro* transcription assays using templates bearing deletions or base substitutions. In virtually all cases examined, a critical 'core domain' was found within ~35 bp of the initiation site. However, efficient expression both *in vitro* and *in vivo* requires an additional **u**pstream **c**ontrol **e**lement (UCE) that extends to around 150 bp upstream of the start site.

Although the distance between the core and upstream domains can be altered, acceptable spacings show the same periodicity as the DNA helix; this suggests that factors bound to the two domains must interact and therefore need to be on the same face of the duplex. This basic organization is conserved, but the actual sequences of rRNA promoters are highly diverged across species. For this reason, rRNA genes can only be transcribed using factors from the same or closely related species. For example, the human pol I machinery will not express a mouse rRNA gene, or vice versa. This situation contrasts strikingly with the pol II and pol III systems, where genes from yeast can often be transcribed using human proteins. The exceptionally high rate of rRNA promoter evolution is thought to occur because the pol I factors need only to recognize one promoter, the many copies of which evolve in unison due to genetic exchanges within the tandem rDNA arrays (unequal crossing-over and gene conversion).

Another unusual feature of the pol I system is that ~12-bp transcription termination elements are located not only at the end of the gene, but also immediately upstream of the promoter. This proximal terminator stimulates expression from the adjacent promoter. One way in which it may do this is by protecting the transcription complex from disruption by polymerase molecules reading through from further upstream. Multiple short enhancer elements are located beyond the upstream terminators, within the spacer regions that separate the rRNA genes. These are interspersed with additional copies of the promoter sequences (Fig. 6.2). Although they are removed from any coding sequences and yield only tiny amounts of stable transcript, the

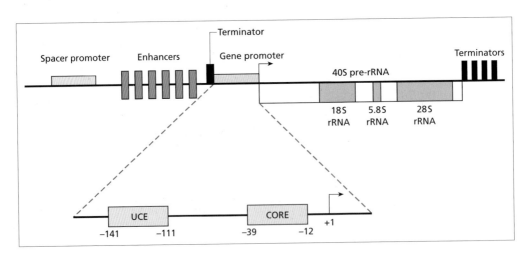

Fig. 6.2 The rRNA transcription unit. The diagram shows the organization of one of the tandem repeats of mouse ribosomal DNA. A similar arrangement is found in humans and in *Xenopus*. The region coding for the 40S pre-rRNA is boxed, with segments encoding the mature rRNAs shaded. The spacer separating the coding regions contain the gene promoter, a terminator, repetitive enhancer elements of ~140 bp and spacer promoters. Below is shown in more detail the gene promoter, with its core region and upstream control element (UCE).

'spacer promoters' can stimulate considerably the level of expression of the downstream gene promoter.

Initiation of pol I transcription

In the vertebrate pol I systems, template recognition is achieved by a factor called UBF—**u**pstream **b**inding **f**actor. UBF is a modular polypeptide of 85–97 kDa, which contains an N-terminal dimerization domain, multiple **h**igh **m**obility **g**roup (HMG) DNA-binding domains, and a C-terminal acidic tail. The tail is composed almost entirely of acidic residues and serines, which are heavily phosphorylated *in vivo* (Fig. 6.3). In both mammals and frogs, UBF is produced as two distinct forms due to alternative splicing. Whereas both forms of *Xenopus* UBF are functional, the shorter form of mammalian UBF is inactive because it is missing the second of its six HMG domains. *Xenopus* UBF has only five HMG domains anyway and consequently cannot function on a mammalian rRNA promoter. When produced individually, each of the *Xenopus* HMG domains can bind to DNA; however, domains 1–3 are essential for transcription, whereas domains 4 and 5 are dispensable.

Despite having multiple DNA-binding domains, UBF shows relatively low sequence specificity *in vitro*. This may reflect the fact that it binds DNA in the minor groove, where bases are less easily discriminated. Nevertheless, UBF manages to recognize sequences in the promoter core, UCE and intergenic enhancers. UBF exists as a dimer in solution and also when bound to DNA. The consecutive HMG boxes interact with one face of the duplex in a colinear manner (Fig. 6.4). A characteristic feature of HMG domains is their ability to bend DNA. UBF provides a striking example of this, because electron-spectroscopic imaging has shown that each dimer can wrap 180 bp of DNA around itself in a single turn of approximately 360°.

The three-dimensional nucleoprotein complex assembled by UBF is recognized by a factor with many different names. The human version is generally referred to as SL1, whereas the mouse factor has been called TIF-IB and the *Xenopus* factor Rib1. Although confusing, the various names serve to remind us that these proteins are not interchangeable between species. Indeed,

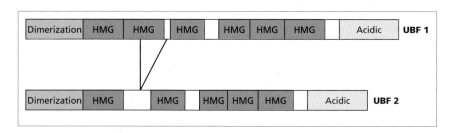

Fig. 6.3 The domain structure of mammalian UBF. UBF2 is inactive, because 37 residues encoded by exon 8 are spliced out.

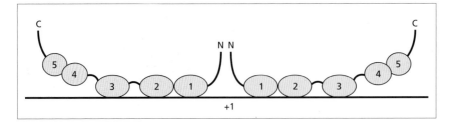

Fig. 6.4 The HMG domains of UBF contact DNA in a colinear manner. The diagram represents a *Xenopus* UBF dimer interacting with the core promoter region.

SL1 (human)	TIF-IB (mouse)
TAF$_I$110	TAF$_I$95
TAF$_I$63	TAF$_I$68
TAF$_I$48	TAF$_I$48
TBP	TBP

Table 6.1 Subunit composition of human SL1 and mouse TIF-IB.

whereas SL1 is dependent upon UBF for recruitment to the promoter, TIF-IB is able to bind to the promoter and support a low level of transcription in the absence of UBF. Both SL1 and TIF-IB have been purified to homogeneity and shown to consist of four subunits (Table 6.1). In both cases, the smallest subunit is TBP (TATA-binding protein). This came as a considerable surprise, because a TATA motif is not part of the rRNA promoter. Indeed, subsequent studies showed that the **TBP-a**ssociated **f**actors (TAFs) present in SL1 prevent TBP from recognizing a TATA motif. The TAFs in SL1/TIF-IB are completely different from those found in TFIID and provide this complex with unique functional properties. Thus, SL1 cannot substitute for TFIID in supporting pol II transcription, and neither can TFIID support pol I transcription in the place of SL1. The TAFs are also responsible for the species specificity of SL1 and TIF-IB, because the least conserved is only 66% identical between man and mouse. The role of TBP in this complex remains unclear. However, its presence raised an important question: what prevents the assembly of TBP-containing complexes involving inappropriate mixtures of TAFs from different systems? This is an important issue, because a complex composed of a heterologous mixture of TAF$_I$s and TAF$_{II}$s would almost certainly be inactive and might interfere with the function of bona fide SL1 and TFIID. Reconstitution experiments carried out with recombinant subunits showed that once one TAF$_I$ has bound to TBP, it can prevent the recruitment of TAF$_{II}$250, TAF$_{II}$150 or TFIIB. Conversely, prior binding of TAF$_{II}$250 or TAF$_{II}$150 prevents TBP from interacting with the TAF$_I$s (Fig. 6.5).

Once UBF and SL1/TIF-IB/Rib1 are in position, pol I can enter the complex (Fig. 6.6). In the murine system, this has been shown to involve a direct inter-

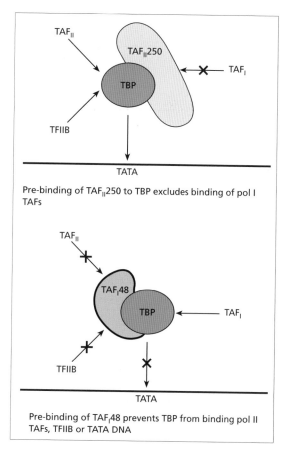

Pre-binding of TAF$_{II}$250 to TBP excludes binding of pol I
TAFs

Pre-binding of TAF$_I$48 prevents TBP from binding pol II
TAFs, TFIIB or TATA DNA

Fig. 6.5 Complexes containing a mixture of TAF$_I$s and TAF$_{II}$s do not form. This must be very important, as heterogeneous mixtures are unlikely to be functional. Binding of TAF$_{II}$250 to TBP precludes the subsequent recruitment of TAF$_I$s, but allows TFIIB and other TAF$_{II}$s into the complex. Conversely, binding of TAF$_I$48 excludes TAF$_{II}$s and TFIIB but allows the recruitment of other TAF$_I$s. It will be intriguing to learn the structural basis of these effects.

action between the HMG domains of UBF and a unique subunit of pol I. Just as pol II is recruited to class II genes as a preformed complex with TFIIF, mouse pol I brings with it two additional factors called TIF-IA and TIF-IC. Neither of these pol I-associated factors have been fully characterized, although both appear important for rRNA synthesis in mice. TIF-IA activity responds to extracellular growth conditions and thereby helps ensure that rRNA production is appropriate to meet demands; if growth factors are limiting, TIF-IA becomes inactivated and pol I transcription is downregulated. TIF-IC performs several roles: it is required for the assembly of a stable initiation complex and also for the formation of the first internucleotide bonds. In addition, TIF-IC stimulates the overall rate of transcript elongation and suppresses pausing by pol I. In these respects, TIF-IC is strongly reminiscent of TFIIF; both are polymerase-associated factors and both operate during initiation and elongation. Given these functional similarities, it will be extremely interesting to see whether TIF-IC also bears homology to TFIIF at the sequence and/or structural levels. TIF-IC is reported to have a molecular mass of 65 kDa, but as yet nothing more is known about its composition.

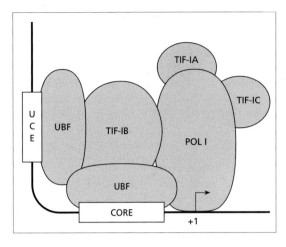

Fig. 6.6 Interactions between the murine pol I factors in a preinitiation complex. UBF binds to both the core and upstream control element (UCE) of the promoter and interacts directly with pol I and TIF-IB. TIF-IA and TIF-IC interact with pol I.

The promoter elements and transcription factors of the pol I system in yeast bear little or no sequence homology to those of higher organisms. Nevertheless, there are clear similarities in the organization of the systems. The rDNA promoter in *Saccharomyces* consists of an upstream element between about −50 and −155, and a ~50-nucleotide core element that includes the initiation site and extends to −40. The upstream element is recognized by a complex called **u**pstream **a**ctivation **f**actor (UAF), which then recruits **c**ore **f**actor (CF), which in turn recruits pol I. Yeast pol I also has an associated factor, called Rrn3p, which is important for transcription. In all of these respects, the yeast system resembles the vertebrate pol I systems. However, yeast UAF, unlike vertebrate UBF, is a complex of five different polypeptides. Furthermore, TBP does not appear to be an integral component of yeast CF, which is a complex of three polypeptides. Indeed, a substantial proportion of TBP in *Saccharomyces* appears to exist in the absence of TAFs, whereas most if not all of the TBP in higher organisms seems to be assembled into complexes such as TFIID and SL1. This feature greatly facilitated the purification of TBP from yeast, whereas its fractionation from higher systems was considerably complicated by its incorporation into several distinct complexes. For this reason, TBP was first purified from *Saccharomyces*. Although TBP is not part of the yeast CF complex, it is nevertheless involved in assembling CF onto the promoter core. UAF has affinity for TBP, which in turn binds to CF. UAF also makes direct contact with CF, in order to facilitate its recruitment (Fig. 6.7). This is again reminiscent of the higher systems, as human UBF contacts both TBP and TAF$_I$48 in order to recruit SL1.

Once pol I and its associated factors have been assembled at the promoter, transcription can initiate. There is no evidence for a distinct promoter clearance step, as occurs with pol II. UBF and SL1 remain at the promoter as the polymerase moves off, which must greatly accelerate the rate of promoter usage by avoiding the need to assemble a new preinitiation complex for

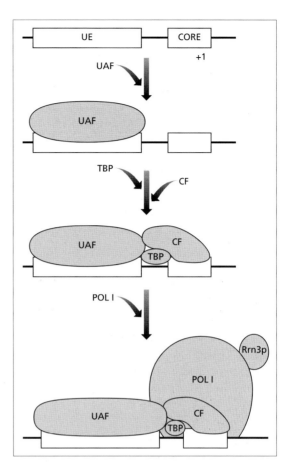

Fig. 6.7 Schematic model of the assembly of a pol I transcription complex in *Saccharomyces cerevisiae*. Promoter recognition is achieved by the upstream activation factor (UAF), which binds to the upstream element. The core factor (CF) is then recruited to the promoter core in a step that involves direct contact between UAF and CF, but also involves TBP acting as an intermediary between these two factors. Once CF is in position, pol I can join along with a polymerase-associated factor called Rrn3p. Additional uncharacterized factors may also be involved.

every round of transcription. Transcript elongation is very efficient, and pol I will synthesize approximately 20–30 nucleotides of rRNA per second. Under conditions of active expression, electron micrographs show a polymerase density of approximately one every 100 bp. These numbers indicate that initiation and termination must occur at least once every 5 s on an rRNA gene *in vivo*.

Termination of pol I transcription

Termination of pol I transcription is dependent on a factor that binds to a specific sequence at the end of the rRNA coding region. In mammals this factor is called TTF-I and in yeast it is called Reb1p. These terminator proteins are significantly related in their DNA-binding domains; both contain two copies of an ~80-residue motif which is referred to as the SANT domain. Furthermore, TTF-I has been shown to terminate transcription by yeast pol I. Thus, it appears that the mechanism for terminating pol I transcription has been

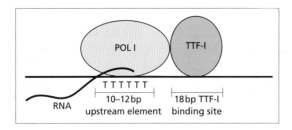

Fig. 6.8 The pol I termination apparatus in mice. Transcription terminates at a T-rich region just upstream of the binding site for TTF-I.

conserved through evolution, in striking contrast to the initiation apparatus. The binding sites for TTF-I and Reb1p are sensitive to orientation and will only terminate transcription coming from upstream. The polymerase stops at an element some 12–20 bp away from TTF-I or Reb1p and the sequence of this element is important for efficient termination (Fig. 6.8). Mutagenesis experiments with the mouse system have shown that the binding of TTF-I to its recognition site is sufficient to cause pol I to pause, but the T-rich upstream element is necessary for transcript release. TTF-I and Reb1p probably have an active role in stimulating the release process, because release is very inefficient if a heterologous DNA-binding protein such as lac repressor is positioned in their place. It is therefore unlikely that TTF-I and Reb1p simply serve as transcriptional road blocks.

TTF-I and Reb1p are large, multidomain proteins that perform additional functions besides their role in pol I termination. Both have been shown to rearrange the nucleosomes in the vicinity of their binding sites and in both cases this activity requires an N-terminal domain that is distinct from the DNA-binding domain. Remodelling the chromatin around the rRNA gene may be important in allowing the initiation factors to gain access to the promoter. In all vertebrates examined, there is a binding site for a pol I terminator protein just upstream of the rRNA gene promoter, in addition to the sites at the end of the coding region. This promoter-proximal site prevents read-through transcription from the upstream spacer promoters. However, it probably also ensures that the chromatin structure of the gene promoter remains open and accessible. Binding sites for Reb1p are found in several pol II promoters, where it stimulates transcription, presumably by rearranging nucleosomes. Indeed, the lethal effects of inactivating Reb1p cannot be overcome by expressing rRNA from a pol II promoter; this demonstrates that Reb1p performs an essential function besides its role in rRNA synthesis.

Apart from terminating transcription and remodelling chromatin, TTF-I also seems to provide a barrier to DNA replication. In both yeast and mammals, DNA replication originates in the intergenic spacer of the rDNA and then spreads bidirectionally. When a replication fork approaches an active rRNA gene from downstream, it encounters a barrier that prevents it from colliding with pol I that is engaged in transcript elongation. This barrier has been mapped to the second of the eight TTF-1-binding sites that follow the mouse rRNA transcription unit. Sequences that flank this site are also important, which suggests that another DNA-binding protein may help block the replication fork. Just as its function in transcription termination is orientation dependent, TTF-1 will only serve as a replication barrier if approached from downstream. As yet, there is no evidence that Reb1p performs a comparable function in yeast.

Further reading

Reviews

Guarente, L. (1997) Link between ageing and the nucleolus. *Genes Dev.* **11**: 2449–2455.

Heix, J. & Grummt, I. (1995) Species specificity of transcription by RNA polymerase I. *Curr. Opin. Genet. Dev.* **5**: 652–656.

Reeder, R. H. (1990) rRNA synthesis in the nucleolus. *Trends Genet.* **6**: 390–395.

Reeder, R. H. & Lang, W. H. (1997) Terminating transcription in eukaryotes: lessons learned from RNA polymerase I. *Trends Biochem. Sci.* **22**: 473–477.

Reeder, R. H., Pikaard, C. S. & McStay, B. (1995) UBF, an architectural element for RNA polymerase I promoters. *Nucl. Acids Mol. Biol.* **9**: 251–263.

Sollner-Webb, B. & Mougey, E. B. (1991) News from the nucleolus: rRNA gene expression. *Trends Biochem. Sci.* **16**: 58–62.

Sollner-Webb, B. & Tower, J. (1986) Transcription of cloned eukaryotic ribosomal RNA genes. *Annu. Rev. Biochem.* **55**: 801–830.

Paule, M. R. & White, R. J. (2000) Transcription by RNA polymerases I and III. *Nucleic Acids Res.* **28**: 1283–1298.

Selected papers

The interaction of UBF with DNA

Bazett-Jones, D. P., Leblanc, B., Herfort, M. & Moss, T. (1994) Short-range DNA looping by the *Xenopus* HMG-box transcription factor, xUBF. *Science* **264**: 1134–1137.

Pol I TAFs

Beckmann, H., Chen, J.-L., O'Brien, T. & Tjian, R. (1995) Coactivator and promoter-selective properties of RNA polymerase I TAFs. *Science* **270**: 1506–1509.

Comai, L., Zomerdijk, J. C. B. M., Beckmann, H., Zhou, S., Admon, A. & Tjian, R. (1994) Reconstitution of transcription factor SL1: exclusive binding of TBP by SL1 or TFIID subunits. *Science* **266**: 1966–1972.

Polymerase-associated factors

Schnapp, A., Pfleiderer, C., Rosenbauer, H. & Grummt, I. (1990) A growth-dependent transcription initiation factor (TIF-IA) interacting with RNA polymerase I regulates mouse ribosomal RNA synthesis. *EMBO J.* **9**: 2857–2863.

Schnapp, G., Schnapp, A., Rosenbauer, H. & Grummt, I. (1994) TIF-IC, a factor involved in both transcription initiation and elongation of RNA polymerase I. *EMBO J.* **13**: 4028–4035.

Assembly of the pol I initiation complex in vertebrates

Bell, S. P., Learned, R. M., Jantzen, H.-M. & Tjian, R. (1988) Functional cooperativity between transcription factors UBF1 and SL1 mediates human ribosomal RNA synthesis. *Science* **241**: 1192–1197.

McStay, B., Hu, C. H., Pikaard, C. S. & Reeder, R. H. (1991) xUBF and Rib1 are both required for formation of a stable polymerase I promoter complex in *X. laevis*. *EMBO J.* **10**: 2297–2303.

The pol I transcription apparatus in yeast

Steffan, J. S., Keys, D. A., Dodd, J. A. & Nomura, M. (1996) The role of TBP in rDNA transcription by RNA polymerase I in *Saccharomyces cerevisiae*: TBP is required for upstream activation factor-dependent recruitment of core factor. *Genes Dev.* **10**: 2551–2563.

Species specificity

Bell, S. P., Jantzen, H.-M. & Tjian, R. (1990) Assembly of alternative multiprotein complexes directs rRNA promoter selectivity. *Genes Dev.* **4**: 943–954.

Rudloff, U., Eberhard, D., Tora, L., Stunnenberg, H. & Grummt, I. (1994) TBP-associated factors interact with DNA and govern species specificity of RNA polymerase I transcription. *EMBO J.* **13**: 2611–2616.

TTF-I

Gerber, J.-K., Gogel, E., Berger, C. *et al.* (1997) Termination of mammalian rDNA replication: polar arrest of replication fork movement by transcription termination factor TTF-I. *Cell* **90**: 559–567.

Kuhn, A. & Grummt, I. (1990) Specific interactions of the murine termination factor TTF-I with class-I RNA polymerases. *Nature* **344**: 559–562.

Langst, G., Blank, T. A., Becker, P. B. & Grummt, I. (1997) RNA polymerase I transcription on nucleosomal templates: the transcription termination factor TTF-I induces chromatin remodeling and relieves transcriptional repression. *EMBO J.* **16**: 760–768.

Chapter 7: Transcription by RNA Polymerase III

Pol III is the largest and most complex of the RNA polymerases, having 17 subunits in both humans and yeast and an aggregate mass of 600–700 kDa. The genes transcribed by pol III encode a variety of small RNA molecules which do not get translated (Table 7.1). Many of these have essential functions in cellular metabolism, such as tRNA and 5S rRNA, which are required for protein synthesis, 7SL RNA, which is involved in intracellular protein transport as part of the signal recognition particle, and the U6, H1 and MRP RNAs, which are involved in the processing of RNA transcripts. The VA RNAs encoded by adenovirus are also synthesized by pol III, and these serve to divert the translational machinery of an infected cell towards the more effective production of viral proteins. Other class III genes encode transcripts with no known function. This category includes the 7SK genes and the short interspersed repeat (SINE) gene families, such as Alu, which constitute the majority of pol III templates in mammals. Overall, pol III is reckoned to be responsible for ~10% of nuclear transcription. However, this figure is dependent on the growth state of cells, because rapidly growing cells synthesize pol III transcripts much more actively than quiescent ones; the reason for this linkage is presumed to be the same as for pol I, namely that the rate of growth is directly proportional to the rate of accumulation of protein, which is clearly dependent upon adequate supplies of tRNA and rRNA. Like pol I, pol III transcription is abnormally elevated in a broad range of transformed and tumour cell types, in keeping with the deregulation of growth and proliferation that accompanies carcinogenesis.

Table 7.1 Pol III products.

Product	Function
tRNA	Translational adaptor
5S rRNA	Ribosomal component
U6 snRNA	mRNA splicing
H1 RNA	RNase P component (tRNA processing)
MRP RNA	rRNA splicing
7SL RNA	Signal recognition particle component
7SK RNA	Unknown
SINE transcripts	Unknown
VA RNA	Translation control (adenovirus)

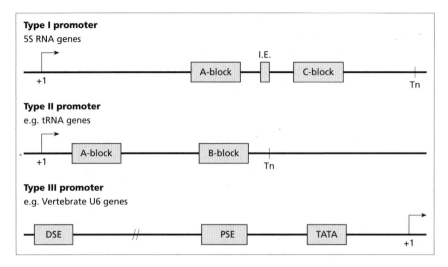

Fig. 7.1 The three types of promoter arrangement that are utilized by pol III. The site of transcription initiation is indicated by +1 and the site of termination is indicated by Tn. Promoter elements are shown as open boxes. IE stands for intermediate element.

Promoter structure

The most striking and unusual feature of the promoters used by pol III is that the majority include important sequence elements downstream of the transcription start site, within the transcribed region. These internal control regions are generally discontinuous structures, composed of essential blocks separated by non-essential regions. A classic example is provided by the somatic 5S rRNA genes of *Xenopus laevis*, which require three internal elements for efficient transcription; an A-block located between +50 and +64, an intermediate element at +67 to +72, and a C-block from +80 to +97 (Fig. 7.1). The promoter is relatively intolerant of changes in the spacing between individual elements. This type of internal control region is also found in the 5S rRNA genes of many other organisms. However, it is a peculiarity of the 5S rRNA genes and is often referred to as a type 1 promoter.

The most common promoter arrangement that is used by pol III is found in the tRNA genes, as well as the adenovirus VA genes and many SINE gene families. It is called a type II promoter and consists of two highly conserved sequence blocks, A- and B-, located within the transcribed region (Fig. 7.1). The A-blocks of type I and II promoters are homologous and sometimes interchangeable, but in the latter case they are much closer to the start site. For example, a *Xenopus* tRNA^{Leu} gene has its A-block between +11 and +21, ~40 bp further upstream than the A-block of the *Xenopus* 5S genes. The location of the B-block is extremely variable, partly because some tRNA genes have short introns within their coding regions. Interblock separations of ~30–60 bp are optimal for transcription, although a distance of 365 bp can be tolerated. The remarkable thing about this flexibility is that the A- and B-blocks are bound simultaneously by a single factor, called TFIIIC.

The first eukaryotic promoters to be characterized belonged to tRNA and 5S rRNA genes. This created an initial impression that higher organisms utilize an entirely different promoter organization from the upstream regulatory sequences that had been mapped in prokaryotes. Subsequent analyses revealed, of course, that internal promoters are exceptional and that upstream control elements are the norm for eukaryotic pols I and II, just as in bacteria. Indeed, a few vertebrate class III genes lack any requirement for intragenic promoter elements; these are referred to as type III. For example, human and mouse U6 snRNA promoters have been identified which retain full activity following deletion of all sequences downstream of the initiation site. It is very peculiar that an extragenic promoter organization appears to have evolved relatively recently within the pol III system. In yeast, U6 genes have functional A- and B-blocks, albeit in unusual positions. A U6 gene with an entirely internal promoter has also been found in humans.

The best characterized type III promoter belongs to a human U6 gene (Fig. 7.1). The sequences required for efficient expression are a TATA box, located between –30 and –25, a **p**roximal **s**equence **e**lement (PSE) between –66 and –46, and a **d**istal **s**equence **e**lement (DSE) between –244 and –214. The U6 PSE and DSE are homologous and interchangeable with elements found at comparable positions in the U2 snRNA gene that is transcribed by pol II. However, a TATA box is not found in the U2 promoter; this is another curious anomaly, because TATA sequences are a classic feature of class II rather than class III genes. Even more paradoxical is the observation that inserting a TATA box can convert U2 into a pol III promoter, whereas crippling its TATA box allows U6 to be transcribed by pol II. Clearly, the U snRNA genes are a law unto themselves!

Transcription complex assembly on type I and II promoters

The A- and B-block sequences found in type II promoters are recognized by the multisubunit complex TFIIIC. This is one of the largest and most complicated transcription factors that has been studied. In *Saccharomyces*, TFIIIC consists of two globular domains, each of ~300 kDa and ~10 nm in diameter. It is composed of six subunits (Table 7.2), none of which seems able to bind specifically to DNA on its own. The photocrosslinking technique (Box 7.1) was used to map where the various subunits of TFIIIC are located when bound to a promoter. They were found to extend across the entire length of a tRNA gene (Fig. 7.2). Both the A- and B-blocks are contacted, although the latter is the major determinant of binding affinity. It is remarkable that these two promoter elements are recognized simultaneously by a single factor, because their separations can vary substantially between different genes. Furthermore, the

Table 7.2 TFIIIC subunits in *Saccharomyces cerevisiae*.

Subunit	Gene	Features
138 kDa	*TFC3*	Two regions of weak HMG homology
131 kDa	*TFC4*	Multiple TPR repeats; HLH homology
95 kDa	*TFC1*	HTH homology
91 kDa	*TFC6*	No significant homologies
60 kDa		Not yet reported
55 kDa	*TFC7*	Chimeric protein generated by chromosomal rearrangement

Box 7.1 Photocrosslinking

The technique of photocrosslinking is used to identify which polypeptides bind to a particular region of DNA. A DNA fragment is prepared that contains one or more radioactive nucleotides and also one or more photosensitive nucleotides. Many different types of photosensitive nucleotides are available, but commonly used examples are bromodeoxyuridine (BrdU) and [N-(p-azidobenzoyl)-3-aminoallyl]-deoxyuridine (N$_3$RdU), both of which are incorporated in place of thymidine. The photoreactive DNA probe is incubated with proteins under conditions that allow DNA recognition. The sample is then irradiated with UV light, which converts the photosensitive nucleotide into an unstable and highly reactive intermediate that rapidly forms a covalent bond with any nearby protein, thereby crosslinking it to the DNA. For example, the azide group of N$_3$RdU is converted to a highly reactive nitrene. After crosslinking, an excess of DNase I is added, which digests away any probe that is not protected by crosslinked protein. The sample is then run on an SDS–polyacrylamide gel to separate the proteins according to size. The gel is dried and exposed to X-ray film, in order to identify the molecular mass of polypeptides which have been 'tagged' by crosslinking to radioactive DNA. For example, when crosslinking reactions were carried out with human TFIIIC2 and a BrdU-substituted VA gene probe, only the 220-kDa subunit of TFIIIC2 was labelled, indicating that this is the DNA-binding subunit. More sensitive studies were carried out using yeast TFIIIC and a tRNA gene that had been substituted at specific positions with N$_3$RdU. Whereas BrdU only crosslinks proteins that are in close contact with the DNA, the reactive group of N$_3$RdU is on a long 1 nm tether that protrudes out of the major groove, allowing the space around the DNA to be probed. By using a series of probes in which N$_3$RdU was introduced at different positions throughout the promoter of tRNA and 5S rRNA genes, it was possible to map the locations of each of the subunits of TFIIIC along the DNA. The results of these experiments are summarized in Fig. 7.2.

Box 7.1 (contd.)

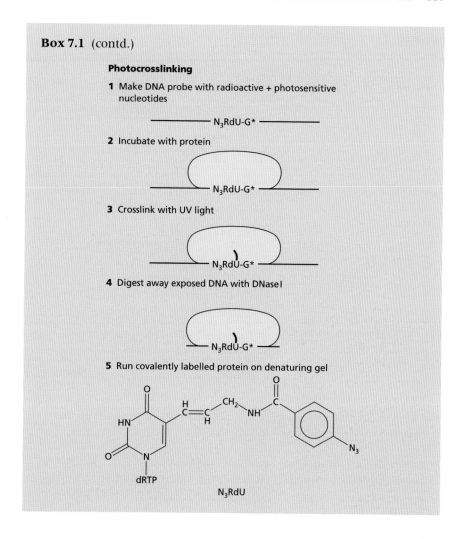

Photocrosslinking

1 Make DNA probe with radioactive + photosensitive nucleotides

2 Incubate with protein

3 Crosslink with UV light

4 Digest away exposed DNA with DNaseI

5 Run covalently labelled protein on denaturing gel

relative helical orientation of the A- and B-blocks is not important for transcription efficiency. Electron-microscopic analysis suggests that a linker region between the two domains of TFIIIC can stretch, giving the protein the appearance of a dumbell when bound to large tRNA genes (Fig. 7.3). However, on promoters with very long interblock separations the ability to stretch is exceeded and the intervening DNA is looped out. This capacity of TFIIIC to accommodate such a diversity of promoter sizes constitutes an unprecedented feat of protein flexibility; crystallographic analysis of the details will prove fascinating.

Human TFIIIC seems to be rather different from the yeast factor. It can be resolved by ion-exchange chromatography into two components, called TFIIIC1 and TFIIIC2. Whereas both components are required for expression of 5S rRNA, VA and tRNA genes, U6 and 7SK transcription requires TFIIIC1 but not TFIIIC2. The initial recognition of type II promoters is achieved by TFIIIC2, which then serves to recruit TFIIIC1 and TFIIIB. TFIIIC1 enhances

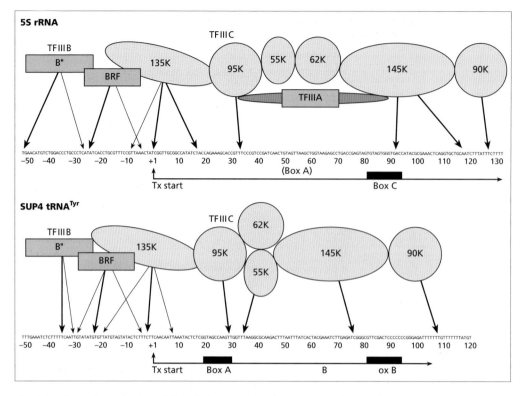

Fig. 7.2 Schematic comparison of the relative positions of the various subunits of *S. cerevisiae* TFIIIC when bound to tRNA or 5S rRNA genes.

and extends the footprint produced by TFIIIC2. Little progress has been made in the characterization of TFIIIC1, although sedimentation analysis suggests that it has a mass of up to 200 kDa. Human TFIIIC2 has been purified and shown to consist of five polypeptides, of 220, 110, 102, 90 and 63 kDa, giving a cumulative mass approaching 600 kDa. The largest subunit is responsible for binding DNA; however, it displays no significant homology to any of the subunits of *Saccharomyces* TFIIIC, which is very surprising because the A- and B-blocks are well conserved between mammals and yeast.

Productive recruitment of TFIIIC to 5S rRNA promoters requires the presence of the gene-specific factor TFIIIA. *Xenopus* TFIIIA was the first eukaryotic transcription factor to be purified to homogeneity and the first to have its cDNA cloned. It was also the founder member of the zinc finger family, because most of its 344 amino acid residues are taken up by nine tandem, zinc-dependent DNA-binding domains that are referred to as fingers (Fig. 7.4). An X-ray crystal structure has been published showing six TFIIIA fingers bound to a 5S rRNA gene (Fig. 7.5). The C-block is recognized by the N-terminal three fingers (fingers 1–3), which contribute ~95% of the total binding energy of full-length TFIIIA; these fingers wrap smoothly around the major groove in a manner that is typical of this class of DNA-binding domain. Fingers 7–9 are thought to contact the A-block in a similar fashion, but with lower affinity.

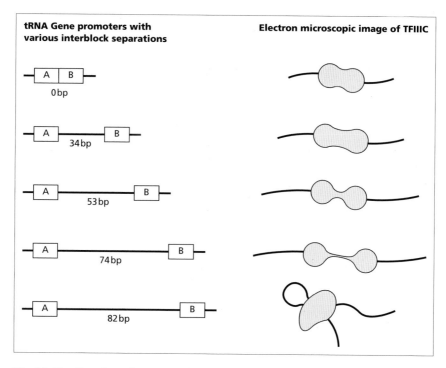

Fig. 7.3 The distortions of *S. cerevisiae* TFIIIC when bound to the promoters of tRNA^{Leu} genes with various interblock separations, as revealed by scanning transmission electron microscopy. When the A- and B-blocks are immediately adjacent (a), TFIIIC appears as a single large mass. As the separation is increased to 34 bp (b), 53 bp (c) or 74 bp (d), the two DNA-binding domains are stretched further apart, until TFIIIC has the appearance of a dumbell. When the interblock separation is increased to 82 bp (e), TFIIIC can stretch no further; its two domains snap back together and the DNA between the A- and B-blocks is looped out. Reproduced with permission from Schultz, The two DNA-binding domains of yeast transcription factor as observed by scanning transmission electron microscopy, *EMBOL* **8** 3815–3824. Copyright (1989) Oxford University Press.

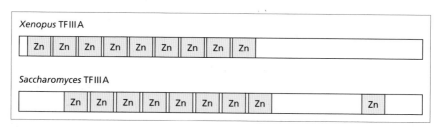

Fig. 7.4 Locations of the zinc fingers within the primary structures of TFIIIA from *Xenopus laevis* and *Saccharomyces cerevisiae*.

The middle three fingers adopt a completely different configuration in order to span the interblock DNA, which is twice as long as the regions bound by fingers 1–3 or 7–9. They run along one side of the duplex in an open, extended structure. Of these, only finger 5 makes base contacts in the major groove, at the intermediate element, whereas fingers 4 and 6 straddle the neighbouring

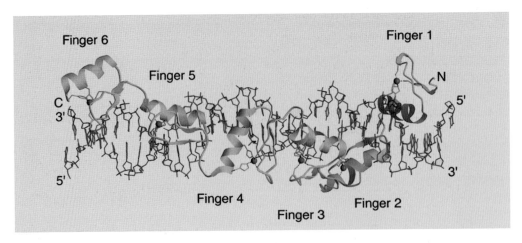

Finger 6

Finger 1

Finger 5

C
3'

N

5'

5'

3'

Finger 4

Finger 2

Finger 3

Fig. 7.5 Model illustrating how the nine C_2–H_2 fingers of *Xenopus* TFIIIA are thought to track along the DNA. The positions of fingers 1–6 have been visualized directly by X-ray crystallography, whereas the locations of fingers 7–9 have been inferred from biochemical analyses. Whereas most of the fingers follow the major groove, fingers 4 and 6 straddle the minor groove allowing a greater distance to be covered. In this way, TFIIIA is able to make contacts that stretch over 55 bp. Reproduced with permission from Nolte, Differing roles of zinc fingers in DNA recognition, *PNAS* **95**, pp. 2938–2943. Copyright (1998) National Academy of Sciences, USA.

minor grooves and function primarily as spacer elements. In this way, TFIIIA is able to occupy an unusually extended site; this 38-kDa protein makes regular contacts across 50 bp of DNA, which contrasts with the more typical behaviour shown by the 100-kDa polypeptide Oct-1 that interacts with an 8-bp sequence motif.

TFIIIA also has nine zinc fingers in *S. cerevisiae*. However, the spacings are very different from *Xenopus* TFIIIA and overall the two proteins are only ~20% identical, with most of the shared residues occurring in the finger motifs (Fig. 7.4). Even among frog species there is considerable divergence, with TFIIIA from *X. laevis* showing only 84% identity to the *X. borealis* protein and 63% identity to the *Rana catesbeiana* protein. This rapid evolution of TFIIIA, which is dedicated to the metabolism of 5S rRNA, is consistent with the lack of conservation of the pol I factors that are responsible for producing the other rRNA species.

TFIIIA serves as an adaptor, providing a platform that allows TFIIIC to be recruited onto 5S rRNA genes for which it has little affinity. Photocrosslinking has shown that the locations of the various TFIIIC subunits, relative to each other and to the initiation site, are very similar on a yeast 5S rRNA gene to those seen on a tRNA gene (Fig. 7.2). This is remarkable, given the dissimilarity in promoter structures and the additional presence of TFIIIA. It is also notable that although tRNA and 5S rRNA genes both employ an A-block, it is bound by TFIIIC in the former case but by TFIIIA in the latter.

Regardless of the promoter type, the primary function of TFIIIC seems to be in recruiting an essential initiation factor called TFIIIB. This process has been characterized extensively in yeast, where *Saccharomyces* TFIIIB has been reconstituted entirely from recombinant components. It is a complex of three polypeptides, one of which is **TATA-b**inding **p**rotein (TBP). Thus, TBP is utilized by all three nuclear RNA polymerases. The largest of the TBP-associated factors in TFIIIB is a 90-kDa polypeptide called B″, which displays little homology to other known proteins. This subunit is remarkably resistant to truncation and will continue to support U6 transcription after all but 176 of its 594 residues have been deleted. The other component of *Saccharomyces* TFIIIB is a 70-kDa subunit that displays 23% identity and 44% similarity to TFIIB in its N-terminal 320 residues. Because of this homology, it is often referred to as TFI**IB-R**elated **F**actor or BRF. Given their similarity, it had been expected that BRF and TFIIB might interact with TBP in a similar manner. Indeed, a weak interaction is observed between TBP and the TFIIB-homologous region of BRF. However, the principal TBP-binding domain of BRF maps outside of the region that is conserved with TFIIB. The two separate TBP-binding domains of BRF interact with opposite faces of the TBP–DNA complex. A notable feature of BRF is that it can be split down the middle to give separated halves which continue to function efficiently when recombined. This property has allowed dissection of the roles of the individual domains. Like the yeast protein, human TFIIIB contains TBP and homologues of BRF and B″.

Because it contains TBP, TFIIIB can bind independently to a TATA box. However, most type I and II promoters that are used by pol III lack a TATA sequence and so cannot be recognized directly by TFIIIB. In these cases, TFIIIB is recruited by protein–protein contacts between BRF and DNA-bound TFIIIC. Once recruited, yeast TFIIIB occupies a region of ~40 bp immediately upstream of the transcription start site. From there it is able to bring pol III to the promoter and position it over the initiation region. All three subunits of TFIIIB are required for pol III recruitment, but direct interactions have only been identified in the case of BRF.

Once assembled, transcription complexes on class III genes can display astonishing stability. Even in *Xenopus* erythrocytes, which are transcriptionally silent and bereft of polymerase, complexes remain associated with class III genes for days and perhaps weeks; this was demonstrated by the ability of nuclei isolated from these cells to synthesize tRNA and 5S rRNA when supplemented with purified pol III. The complexes can withstand salt concentrations that would preclude their formation entirely. For example, complexes formed on yeast tRNA and 5S rRNA genes retain full activity after exposure to 500 mM NaCl. In yeast, the interaction between TFIIIB and DNA is the most resistant to salt, whereas TFIIIC and TFIIIA are dissociated more readily. Once recruited to a promoter via TFIIIC, yeast TFIIIB will remain stably bound even in 1 M KCl. This is remarkable, given that TFIIIB alone is incapable of recognizing a TATA-less class III gene. It seems that interaction with TFIIIC unmasks a latent DNA-binding capacity that locks TFIIIB onto a promoter. Indeed, a proteolytic fragment containing the C-terminus of BRF can bind DNA in a sequence-independent fashion, but this cryptic function is not detected with the isolated full-length polypeptide. It has been suggested that interactions between the TFIIIB subunits may contort the DNA and constrain it in such a way that it is unable to slide free; this might allow avid binding in the absence of sequence-specific recognition.

This unusual property of yeast TFIIIB was exploited to show that it is sufficient on its own to recruit pol III and direct multiple rounds of accurately initiated transcription. Thus, TFIIIC and TFIIIA were stripped from fully assembled transcription complexes by exposure to high salt, leaving TFIIIB alone on tRNA and 5S rRNA promoters; the efficiency of transcription was not compromised by this treatment. On this basis, TFIIIA and TFIIIC can be regarded as assembly factors that are dispensable for transcript initiation. It seems highly likely that this conclusion will also apply to metazoan systems, but this has not been tested directly. However, with frog and human factors the dissociation pathway is the reverse of the assembly pathway, polymerase being lost first and then TFIIIB, as the salt concentration is elevated. Thus, the exceptional salt stability of promoter-bound TFIIIB that is seen in simpler organisms appears not to be a feature of vertebrates.

Transcription complex assembly on type III promoters

The type III promoters associated with vertebrate 7SK and U6 snRNA genes have distinct factor requirements from most pol III templates. They utilize TFIIIC1, but not TFIIIC2. Furthermore, the TFIIIB employed by type III promoters is separable chromatographically from the form used by types I and II, probably representing a subcomplex. The PSE of type III promoters is recognized by the factor PTF (**P**SE-binding **T**ranscription **F**actor also called SNAPc), a complex of five subunits which also interacts with the PSE sequences upstream of the U1 and U2 snRNA genes that are transcribed by pol II.

PSE occupancy is of primary importance in assembling a stable preinitiation complex on a type III promoter. The distance separating the PSE and the TATA box is constrained very precisely in type III promoters, apparently because PTF interacts with TATA-bound TBP or TFIIIB. Indeed, PTF and TBP enhance each others recruitment to the U6 promoter. Both PSE and TATA binding are relatively slow, and assembly of the initiation complex takes longer for U6 genes than it does for tRNA or 5S rRNA genes. The long lag phase can be diminished by Oct-1, which binds to the upstream DSE and stimulates occupancy of the PSE. Association of PTF with the PSE of the human U6 gene takes over an hour in the absence of Oct-1, but is complete in 15–30 minutes in its presence. Occupation of the U6 promoter therefore appears to be achieved by a series of cooperative interactions between Oct-1, PTF and TBP. It remains to be determined how TFIIIC1 is recruited, but this may be a late step in complex assembly on type III promoters.

Transcript initiation and elongation

Once pol III has been recruited to a promoter, it melts the DNA helix around the initiation site. This process may require the active participation of TFIIIB, as certain mutations in BRF or B″ allow normal polymerase recruitment and positioning but prevent the formation of a strand-separated open promoter complex. As pol III progresses into the gene, the bubble of melted DNA moves with it. At 20°C, yeast pol III elongates RNA at an average rate of ~20 nucleotides/second, which is similar to the chain elongation rate measured for pol II *in vivo*. However, elongation does not proceed at a uniform rate; for example, at 20°C it takes pol III 3.0 seconds to traverse from nucleotide 17 to nucleotide 46 of a tRNA^{Tyr} gene and 4.1 seconds to travel the next nine nucleotides. This uneven progress results from pausing at internal sites and the rate of extension at individual nucleotides can vary by 31-fold. No elongation factors have been identified for pol III, unlike pols I and II. They may not be necessary because of the extremely small size of class III genes. Alternatively, because pol III has the most subunits of the RNA polymerases, it is possible that one or more of the pol III-specific subunits performs the functions carried out by separable elongation factors in the other systems.

One of the great puzzles of the pol III system is how a class III gene can be expressed when the enormous factor TFIIIC is sitting in the transcribed region. A large complex within the coding region of a gene might be expected to block progression of polymerase or to be displaced by it. However, assembled transcription complexes are not removed from internal control regions by passage of pol III. Furthermore, removing TFIIIC from a yeast tRNA gene made no significant difference to the rate of RNA elongation. During transcription in the normal direction, the presence of TFIIIC delays pol III for just 0.2 seconds at a single site upstream of the B-block. Because the time required for promoter clearance limits initiation rates to below 0.5/second, a downstream delay of 0.2 seconds will make no difference to the overall level of transcription. However, if pol III is engineered artifically to transcribe in the antisense direction and encounter TFIIIC from downstream, it pauses for around 9 seconds before continuing through the B-block. By contrast, TFIIIB prevents the passage of pol III approaching from downstream for over an hour. A full *Xenopus* 5S rRNA gene initiation complex has also been found not to impede transcription of either DNA strand and the complex remains stably bound following multiple polymerase transits. By contrast, a TFIIIA/5S gene complex in the absence of TFIIIB and TFIIIC is dissociated by passage of pol III. Similarly, TFIIIC alone is rapidly displaced from DNA by pol III. It seems likely that the multiple contacts made by a complete transcription complex may be essential for continued integrity. An obvious possibility is that pol III displaces transiently a given factor from its binding site as it transcribes through the gene, but the factor remains stably associated due to protein–protein contacts with other factors bound to DNA sites that are not in the process of being transcribed. The association of TFIIIB with DNA upstream of the transcription start site may be particularly important in preventing TFIIIC and TFIIIA from being released from the template as pol III transcribes through the internal promoter.

It remains to be determined how pol III is able to displace TFIIIC and TFIIIA during chain elongation. One possibility would involve specific protein–protein interactions inducing conformational changes that trigger release of the factors. However, phage RNA polymerases can dislodge TFIIIA, which argues against a specific interaction between pol III and the factors in its path. It may be that the energy consumed in translocation generates sufficient force to displace obstructive proteins. Alternatively, the process of transcription may indirectly weaken the binding of TFIIIA and TFIIIC due to DNA strand separation or the generation of positive supercoils ahead of pol III.

Transcript termination and reinitiation

Whereas pols I and II require accessory factors in order to terminate transcription specifically, pol III can recognize termination sites accurately and efficiently in the apparent absence of other factors. Simple clusters of four or more T residues can serve as terminator signals. Yeast tRNA genes are followed by a cluster of at least six consecutive T residues, in contrast to *Xenopus* class III genes, where T_4 termination signals are common.

After the initial round of transcription, a stable yeast class III preinitiation complex can direct subsequent cycles five- to 10-fold more rapidly than the first. Thus, during multiple round transcription, synthesis of each tRNA molecule takes approximately 35 seconds, whereas initiation of the first transcript can take ~5 minutes (at 22°C). This is because pol III is recycled without being released from the template; as a consequence, the slow initial step of polymerase recruitment is avoided (Fig. 7.6). TFIIIA, TFIIIB and TFIIIC all bend DNA, and this may facilitate internal recycling by bringing the two ends of a class III gene into close proximity. Human pol III is also retained in the original transcription complex on VA and tRNA genes without dissociating after each round of synthesis.

TFIIIA as an RNA-binding factor

In addition to its role as a transcription factor, TFIIIA also serves important functions in the storage and transport of 5S rRNA. The molecular mechanisms which allow TFIIIA to bind with specificity to both DNA and RNA have attracted considerable attention. It was initially suggested that the DNA helix of the 5S rRNA gene internal promoter might adopt an A-form conformation and thereby resemble the geometry of RNA rather than a typical B-form DNA duplex. In this way TFIIIA could recognize similar structures in its DNA and RNA binding sites. However, NMR and circular dichroism measurements indicated that DNA fragments containing the internal promoter adopt a structure in solution that differs from the A-form and is irregular and intermediate between

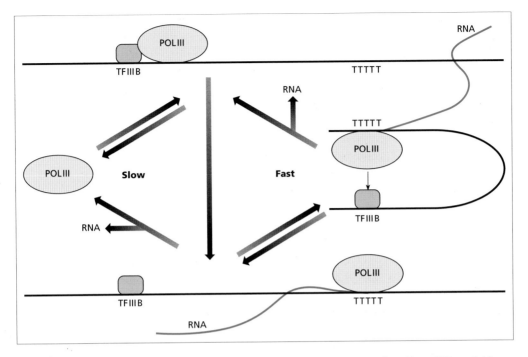

Fig. 7.6 Schematic illustration of the alternative pathways employed by pol III to reinitiate transcription. TTTTT denotes the site of termination.

the classical A- and B-form duplexes, with a wide major groove. Circular dichroism analysis also demonstrated that the binding of TFIIIA does not induce any gross structural transition in the geometry of the promoter DNA.

TFIIIA appears to recognize RNA by a fundamentally different mechanism from that employed in recognizing DNA. The fact that the helices of zinc fingers are too wide to fit into the major groove of an A-form RNA duplex suggests that RNA binding may involve contacts with the phosphate backbone rather than direct recognition of specific nucleotides. In contrast to DNA recognition, the major structural information required for specific RNA binding is provided principally by the secondary and tertiary structure of 5S rRNA and not by its primary sequence. Almost all the RNA molecule is needed to maintain the optimal shape for TFIIIA binding. Whereas fingers 1–3 contribute most of the affinity of DNA binding, fingers 4–7 are of greatest importance in binding RNA. Therefore the RNA-binding and DNA-binding functions of TFIIIA are separable and distinct.

TFIIIA is involved in the nuclear export of 5S rRNA. Two alternative and independent pathways exist for transporting 5S rRNA out of the nucleus in *Xenopus* oocytes; one involves the formation of a 5S ribonucleoprotein (RNP) complex with ribosomal protein L5 and the other is as a 7S RNP complex with TFIIIA. Whereas the 5S RNP shuttles between the nucleus and the cytoplasm, the 7S RNP is retained in the cytoplasm. TFIIIA cannot enter the nucleus if it is bound to 5S rRNA; this may be because the nuclear localization signal of TFIIIA, located within the finger region, is masked in the 7S RNP. Return of TFIIIA to the nucleus therefore requires the exchange of 5S rRNA from the 7S to the 5S RNP. During early oogenesis, 5S rRNA is made in excess over other ribosomal components and is sequestered in the cytoplasm in complex with TFIIIA. As the oocytes develop, L5 levels increase and TFIIIA levels decrease, allowing the stored 5S rRNA to exchange into 5S RNPs and thereby re-enter the nucleus for assembly into ribosomes.

Transport of TFIIIA to the cytoplasm as an RNP complex can deplete the nucleus of this factor, so that it loses the ability to transcribe 5S rRNA genes. This provides a negative feedback loop for regulating the production of 5S rRNA. A simpler version of this feedback inhibition is also observed *in vitro*, where transcription of 5S genes can be inhibited by the presence of excess exogeneous 5S rRNA. Following mutation of finger 6, which plays a primary role in RNA binding but only a subsidiary role in DNA binding, TFIIIA directs transcription that is less susceptible to feedback inhibition. As a consequence, the finger 6 mutant activates 5S rRNA synthesis more efficiently than wild-type TFIIIA in microinjected *Xenopus* embryos.

Coordination of nuclear RNA polymerase activity

An important question that has been largely neglected is how the activities of the three nuclear RNA polymerases are coordinated with respect to each other. Such regulation could be of considerable significance, because a lack of balance might place a huge unnecessary burden on the metabolic economy of the cell. Ribosome synthesis requires equimolar amount of ~80 ribosomal proteins and 28S, 18S, 5.8S and 5S rRNAs. One might anticipate that the genes encoding these products would be controlled coordinately. To a considerable extent this is the case, and the production of ribosomal components is well balanced. In particular, the rates of ribosome biogenesis and transcription by pols I and III are closely coupled with cellular growth. However, the co-ordination of these activities is far from absolute. The levels of 5S rRNA tend to be controlled less precisely than large rRNA and ribosomal proteins, with an excess of up to 20% detected in many cells. When yeast are starved of nutrients pol I transcription declines rapidly, whereas 5S rRNA and then tRNA synthesis follow more slowly. Thus, although pols I and III show similar responses to growth conditions, pol I transcription appears to be the most sensitive. A far more extreme example of this is seen during the differentiation of rat L6 myoblasts, where a rapidly proliferating population of cells changes into a non-dividing syncytial population. The rate of pol I transcription and ribosome accumulation is five-fold reduced in the differentiated myotubes, but synthesis of 5S rRNA and ribosomal proteins is unchanged leading to considerable overproduction. The excess of these components is degraded rapidly, thereby preventing the accumulation of a static pool of unassembled ribosomal precursors. The apparently unnecessary synthesis and destruction of such large populations of macromolecules would seem to be an extraordinary example of cellular profligacy.

The obvious way to coordinate transcription by pols I, II and III would be to target shared components. These three RNA polymerases have five common subunits which could, in theory, be used to achieve coregulation. However, no evidence for such control has been reported to date. The other protein that is shared by the three transcription systems is TBP. By targeting TBP, a repressor protein called Dr1 has been shown to coregulate pols II and III. Dr1 is a 19-kDa nuclear phosphoprotein that was isolated from HeLa cells as an inhibitor of pol II transcription. *S. cerevisiae* contains a Dr1 homologue that is 37% identical to its human counterpart and is essential for viability. Dr1 binds directly to TBP and blocks its association with TFIIA and TFIIB, thereby inhibiting pol II transcription. In the TFIIIB complex, BRF contacts TBP at sites that overlap with both the TFIIA- and TFIIB-interaction surfaces. Accordingly, Dr1 can disrupt the association between TBP and BRF, thereby inactivating TFIIIB (Fig. 7.7). Overexpression of Dr1 will repress transcription of a range of human pol III

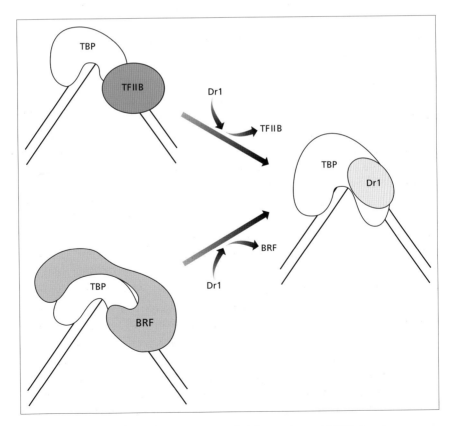

Fig. 7.7 Dr1 can displace TFIIB and BRF from their interactions with TBP, thereby inhibiting transcription by pols II and III.

templates and will also inhibit tRNA synthesis in living yeast cells. By contrast, pol I transcription is not affected by Dr1 in either of these situations. This appears to be because TBP in the SL1 complex is masked from Dr1. Thus, even factors that regulate a shared component may have differential effects. It may be that Dr1 is able to shift the balance of nuclear transcription in favour of pol I. This could be of value under conditions in which the production of large rRNA lags behind the synthesis of other transcripts. It is likely that additional regulatory proteins serve to coordinate the activities of other combinations of polymerases. For example, the retinoblastoma protein RB has been shown to repress pols I and III, but only controls a minority of class II genes. Factors such as these may provide a regulatory network that interlinks the activities of the three nuclear RNA polymerases. Such control may have been extremely important in the evolution of the tripartite eukaryotic transcription system.

Further reading

Reviews

Geiduschek, E. P. & Kassavetis, G. A. (1995) Comparing transcriptional initiation by RNA polymerases I and III. *Curr. Opin. Cell Biol.* **7**: 344–351.

Pieler, T. & Theunissen, O. (1993) TFIIIA: nine fingers–three hands? *Trends Biochem. Sci.* **18**: 226–230.

White, R. J. (1998) *RNA Polymerase III Transcription*. Springer-Verlag, Berlin.

White, R. J. & Jackson, S. P. (1992) The TATA-binding protein: a central role in transcription by RNA polymerases I, II and III. *Trends Genet.* **8**: 284–288.

Willis, I. M. (1993) RNA polymerase III. Genes, factors and transcriptional specificity. *Eur. J. Biochem.* **212**: 1–11.

Selected papers

Polymerase specificity of U6 snRNA genes

Mattaj, I. W., Dathan, N. A., Parry, H. D., Carbon, P. & Krol, A. (1988) Changing the RNA polymerase specificity of U snRNA gene promoters. *Cell* **55**: 435–442.

Electron-microscopic visualization of TFIIIC bound to promoters of varying size

Schultz, P., Marzouki, N., Marck, C., Ruet, A., Oudet, P. & Sentenac, A. (1989) The two DNA-binding domains of yeast transcription factor τ as observed by scanning transmission electron microscopy. *EMBO J.* **8**: 3815–3824.

Use of photocrosslinking to map the topography of a pol III transcription complex

Braun, B. R., Bartholomew, B., Kassavetis, G. A. & Geiduschek, E. P. (1992) Topography of transcription factor complexes on the *Saccharomyces cerevisiae* 5S RNA gene. *J. Mol. Biol.* **228**: 1063–1077.

TFIIIB positions pol III at the transcription start site

Kassavetis, G. A., Braun, B. R., Nguyen, L. H. & Geiduschek, E. P. (1990) *S. cerevisiae* TFIIIB is the transcription initiation factor proper of RNA polymerase III, while TFIIIA and TFIIIC are assembly factors. *Cell* **60**: 235–245.

Passage of the transcribing pol III through factors bound to an internal promoter

Bardeleben, C., Kassavetis, G. A. & Geiduschek, E. P. (1994) Encounters of *Saccharomyces cerevisiae* RNA polymerase III with its transcription factors during RNA chain elongation. *J. Mol. Biol.* **235**: 1193–1205.

Reinitiation by pol III without dissociating from the gene

Dieci, G. & Sentenac, A. (1996) Facilitated recycling pathway for RNA polymerase III. *Cell* **84**: 245–252.

Feedback inhibition of 5S rRNA synthesis

Rollins, M. B., Del Rio, S., Galey, A. L., Setzer, D. R. & Andrews, M. T. (1993) Role of TFIIIA zinc fingers *in vivo*: analysis of single-finger function in developing *Xenopus* embryos. *Mol. Cell. Biol.* **13**: 4776–4783.

Coordinate repression of pols II and III by Dr1

White, R. J., Khoo, B. C.-E., Inostroza, J. A., Reinberg, D. & Jackson, S. P. (1994) The TBP-binding repressor Dr1 differentially regulates RNA polymerases I, II and III. *Science* **266**: 448–450.

Chapter 8: The Influence of Chromatin on Transcription

The human genome would extend for over a metre if unravelled. In order to fit it into a nucleus with a diameter of less than 10 μm, the DNA becomes wrapped up as chromatin and compacted into chromosomes. It is remarkable that genes can remain accessible to transcription factors when in this highly condensed state. Although things that are seldom required may be hidden away, everyday items must be kept somewhere accessible; this is true of chromosomes, which are subdivided into 'open' and 'closed' regions. It is important to bear in mind that the eukaryotic transcription apparatus has evolved to operate in a chromatin environment. Progress in understanding the mechanisms and regulation of transcription will depend upon a clear understanding of the organization and topological state of genes under physiological conditions. The chromatin proteins that serve to compact DNA *in vivo* are not merely a packing material, but provide a dynamic structure that is utilized by the cell to regulate gene expression, sometimes in a complex differential fashion. Indeed, the chromosomal organization of genes can have a profound influence upon their activity. Virtually all the work that has been described in the chapters so far has regarded genes as freely approachable strands of naked DNA. This chapter will consider them in a more physiological condition, organized within the chromosomes, their natural state *in vivo*.

Chromatin structure

Histones are the proteins which are primarily responsible for assembling DNA into chromatin. They are small basic polypeptides of 11–16 kDa which are rich in lysine and arginine residues. They have been well conserved through evolution, as befits their important function in condensing the genetic material. Histones H2A, H2B, H3 and H4 are referred to as the core histones. Each contains a globular domain and a highly charged N-terminal tail that is especially enriched in lysine and arginine. The core histones interact with each other to assemble a structure called the nucleosome, which is the fundamental repeating unit of metazoan chromatin. If a fragmented chromosome is solubilized at very low salt concentrations (< 1 mM NaCl), it appears as a series of beads on a string; the nucleosomes are the beads whereas the DNA is the string. A nucleosome contains a central kernel in which 120 bp of DNA are wrapped around an $(H3/H4)_2$ tetramer. Once the DNA is wrapped around this core histone tetramer, it is joined at each end by a histone H2A/H2B dimer. Recruitment of these dimers depends on both protein–protein interactions with H3 and H4 and on the DNA conformation that is induced by the $(H3/H4)_2$

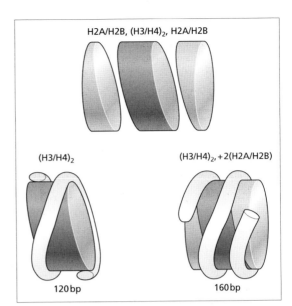

Fig. 8.1 Organization of a nucleosome. A nucleosome consists of a central (H3/H4)$_2$ tetramer, flanked on either side by an H2A/H2B dimer. The tetramer is wrapped by 120 bp of DNA whereas each H2A/H2B dimer organizes an additional 20 bp. Adapted from Wolffe 1995.

tetramer. The resulting octamer of core histones organizes 160 bp of DNA in two left-handed superhelical turns and is referred to as the nucleosome (Fig. 8.1). It is shaped like a disc, with a width of 5.6 nm and a diameter of 11 nm. The globular domains of the individual core histones adopt similar conformations, which are referred to as the histone fold. It consists of long central helix flanked on either side by a loop and then a shorter helix. The central helix serves as an interface for heterodimerization (Fig. 8.2). Three DNA-binding surfaces are generated by dimerization, due to the interaction of loop segments at each end of the long central helix, and through the juxtaposition of the two short α-helices flanking the N-terminal ends of each central helix in the pair. As a result, there is extensive interaction between individual histones in the H3/H4 and H2A/H2B dimers. There is also considerable contact between the H2A/H2B dimers and the (H3/H4)$_2$ tetramer, but this interface is accessible to solvent and is the first to be disrupted when a nucleosome disassembles. The positively charged N-terminal tails of the core histones protrude out of the nucleosome. They are not involved in wrapping DNA around the nucleosome, but interact electrostatically with the phosphodiester backbone. Regulatory modifications of the tails, such as acetylation, may loosen these interactions.

Nucleosomes are generally separated by 20–30 bp that is referred to as linker DNA. In most eukaryotes, this region is associated with a polypeptide called the linker histone, the most common of which is histone H1. However, histone H1 is not found in yeast. Linker histones are slightly larger than the core histones (> 20 kDa) and are highly basic, being very rich in lysine residues. They have a central globular domain and highly charged tails at both the N- and C-termini. Linker histones, such as H1 and H5, contact DNA by means

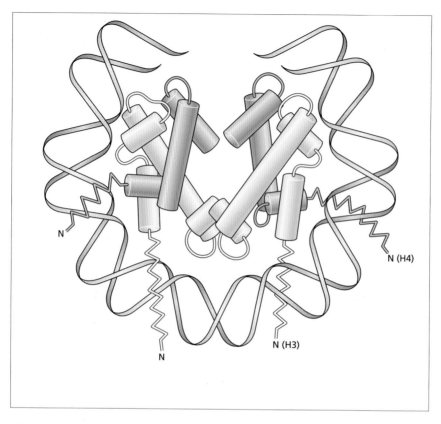

Fig. 8.2 The histone fold, as occurs within an (H3/H4)$_2$ tetramer. Each core histone forms a long central α-helix which is flanked on both sides by a loop and then a shorter α-helix. The helices of H3 and H4 are shown as cylinders. The histone tails are shown as zig-zags, although their precise positioning is unclear. 120 bp of DNA is wrapped around the tetramer. Adapted from Wolffe 1995.

of a structure called the 'winged helix', which is also found in the **h**epatocyte **n**uclear **f**actor **3** (HNF3). This consists of a bundle of three α-helices attached to a three-stranded antiparallel β-sheet. HNF3 utilizes one of the α-helices to bind to DNA across the major groove, and it seems likely that the globular domain of linker histones will contact nucleosomal DNA in a similar fashion. Linker histones only display high affinity for DNA once it has been wrapped around the nucleosome, after which they bind to the adjacent region whilst making protein–protein contacts with the core histones. Association of histone H1 stabilizes the interaction of the nucleosome with the 160 bp of DNA that is wrapped around it. This is likely to involve protein–protein interactions between histones H1 and H2A, as well as DNA contacts by the globular and tail domains of the linker histone. The globular domain of the linker histone contains two DNA-binding sites, which can come into play simultaneously when DNA is wrapped around a nucleosome owing to the juxtaposition of adjacent

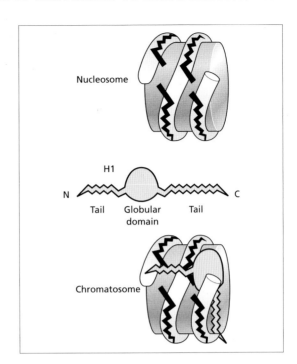

Fig. 8.3 A chromatosome is formed when a single molecule of linker histone, such as histone H1, binds to a nucleosome. The globular domain of the linker histone has two DNA-binding sites which can simultaneously interact with separate regions of the double helix once they are juxtaposed following wrapping around the nucleosome core. The approximate positions of histone tails are indicated by zig-zags. Adapted from Wolffe 1995.

segments of helix. The term chromatosome is used to refer to a nucleosome associated with a linker histone (Fig. 8.3).

The structure of DNA can be severely distorted by assembly into nucleosomes, due to bending and changes in the helical periodicity. Three turns of duplex at the dyad axis of the nucleosome have a helical periodicity of 10.7 bp per turn, which is significantly different from the remainder of the nucleosomal DNA (10.0 bp per turn). The duplex is sharply bent at the junctions between these two types of structure. Because some combinations of bases are more amenable to bending than others, the positioning of nucleosomes is sensitive to DNA sequence.

The ability of transcription factors to recognize their cognate binding sites can be severely restricted by nucleosomes. One face of the helix is occluded because it faces the histone core. A DNA recognition site may therefore be accessible only if it is orientated away from the nucleosome. However, transcription factors that require significantly more than half the helix circumference to bind, because of an extended recognition sequence, may never be able to gain access to the entire motif if it is incorporated into a nucleosome; some part of the recognition sequence may always be occluded, regardless of helical orientation (Fig. 8.4). Furthermore, the adjacent turn of DNA will further restrict access. An additional impediment may be provided by the core histone tails interacting with the phosphodiester backbone. Because DNA in solution has a helical periodicity of 10.5 bp/turn, the overwinding of duplex to 10.0 bp/turn through most of the nucleosome will also impose constraints on

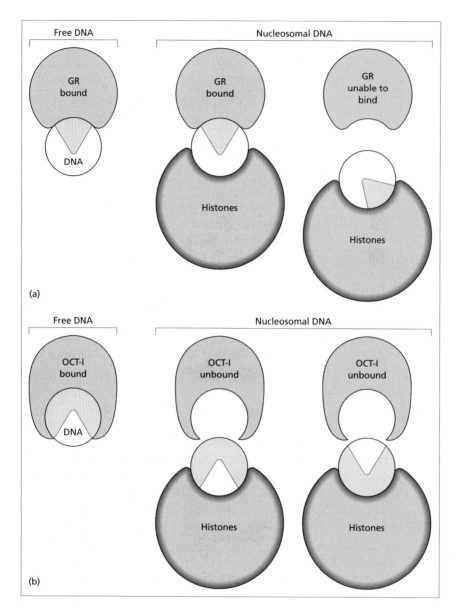

Fig. 8.4 Effect of nucleosomal phase on DNA recognition. Binding of a transcription factor to DNA in a nucleosome can be occluded by the presence of the histone core. Proteins, such as the glucocorticoid receptor (GR), which contact only a narrow sector of the helix (pink) can gain access to nucleosomal DNA if the recognition sequence faces away from the histone core (a). However, proteins that interact with more than half the helical circumference are prevented from binding to nucleosomal sites, regardless of orientation (b). An example is provided by Oct-1, which has an 8-bp recognition motif. A more extreme case is provided by NF1, which has a 17-bp recognition sequence that is extremely inaccessible in nucleosomal DNA.

the base-specific contacts that can be made by a transcription factor. Recognition of extended DNA sequences may also demand that a factor is sufficiently flexible to wrap itself around the histone core.

Linker histones accentuate the accessibility problem presented to transcription factors by encouraging nucleosomal arrays to coil and fold into fibres of chromatin. These fibres are about 30 nm in diameter and are thought to have a solenoidal structure, each turn of which contains six or more nucleosomes and over 1000 bp of DNA. The length of DNA is compacted approximately 50-fold once it is incorporated into nucleosomes and assembled into chromatin fibres. In this condition, large expanses of DNA are likely to be invisible to transcription factors. A key feature of the histones that enables them to achieve this effect is their net positive charge, which allows them to neutralize the phosphodiester backbone, as well as wrapping it around the nucleosomal core. Acetylation of the basic tails may therefore be expected to destabilize chromatin by reducing the charge-balancing effect of the histones. The core histones are sufficient to assemble chromatin fibres, but the solenoids are likely to be relatively dynamic in the absence of linker histones, unfolding with increased frequency. The chromatin fibres are organized further into large domains, typically of 40–100 kb, that are demarcated by non-histone proteins which perform both structural and regulatory roles. These domains undergo further folding within the chromosome, thereby compacting the length of DNA by an additional 100-fold.

Gene repression by nucleosomes

Because the organization and accessibility of DNA in cells is so profoundly affected by histones, one would expect these proteins to have a major impact on gene expression. This prediction is supported by *in vitro* experiments, where the deposition of histones onto naked DNA is generally accompanied by a substantial decrease in transcriptional activity. Nucleosomes have been shown to block the DNA binding of a number of transcription factors by obscuring their recognition sites. A well-characterized example of this effect is provided by the 5S rRNA gene-specific factor TFIIIA. As described in the previous chapter, TFIIIA interacts with approximately 50 bp of promoter DNA, but its high-affinity binding site is the C-block located at the downstream end of the internal control region. If a nucleosome is positioned over the C-block, TFIIIA can no longer bind and 5S rRNA synthesis is repressed. However, if the nucleosome is located further upstream, leaving the C-block open, TFIIIA can still gain access and transcription can proceed, albeit at a reduced rate. Many such positioning experiments were conducted using relatively short fragments of DNA, where the mobility of the nucleosome is constrained by the ends of the template molecule. On longer fragments, it is found that nucleosomes can slide along the DNA, thereby uncovering factor recognition sites. However, this mobility is lost in the presence of the linker histones, which clamp nucleosomes in position, often obscuring important promoter sites.

A striking example of generalized gene repression by histones *in vivo* is provided during the early development of *Xenopus*. These frogs stock-pile many essential products in their oocytes, which allows them to undergo very rapid cycles of cell division following fertilization, without the need for large-scale biosynthesis. Indeed, virtually no transcription occurs during the first 12 cell cycles, until a stage is reached called the **midblastula transition** (MBT), when gene expression resumes. The silencing of transcription during these early rounds of division appears to be mediated by chromatin. A *Xenopus* oocyte contains enough histones to assemble at least 13 000 nuclei. These are sequestered away from the genome until the nuclear envelope breaks down when the oocyte matures. Genes subsequently remain in a repressed state until the MBT. It is thought that by the MBT there has been sufficient synthesis of new DNA to titrate the excess histones. This idea is supported by the observation that injecting exogenous DNA can activate transcription prematurely, before the MBT is reached. This shows that the developing embryo must contain a full complement of transcription factors, in order to be able to express genes when the surplus histones have been titrated away. However, the large excess of nucleosomal proteins prevents these factors from gaining access to promoters. Preincubation of TBP with the *c-myc* promoter allows temporary expression following microinjection, which suggests that TATA binding may be a critical event that is repressed by the histones. Nevertheless, the effect is only transient and the preincubated promoter is soon silenced. It therefore appears that histones present at these very high levels can cause blanket repression of gene expression.

The situation in early *Xenopus* development is clearly highly unusual. In most organisms and under most conditions the production of histones is coordinated with the rate of DNA synthesis. *In vivo* evidence of the influence of histones upon gene expression under more normal conditions has come from genetic experiments carried out in *Saccharomyces cerevisiae*. This involved replacing the chromosomal genes encoding histones with exogenous histone genes in which expression was controlled by a manipulable promoter. When the synthesis of histone H2B or H4 was switched off, several previously silent genes were found to be induced, such as *PHO5* and *CUP1*. These data demonstrate the involvement of the core histones in gene repression *in vivo*.

For many years it was assumed that chromatin serves as a general repressor of gene expression, blanketing down transcription globally in the absence of activators. This idea was entirely plausible, as the whole genome is incorporated into chromatin. Indeed, precisely this appears to occur during the early stages of *Xenopus* development, when the histones are present in vast excess. However, recent data have lead to a more subtle picture of chromatin as a gene-specific regulator, conferring unique patterns of activity upon particular sets of promoters. An extreme and surprising illustration of this was provided by disrupting the linker histone genes of the ciliated protozoan *Tetrahymena thermophila*. As might be expected, the histone H1 knockout strain has enlarged nuclei containing decondensed chromatin. However, these cells grow normally and display no change in the synthesis of rRNA, tRNA or the bulk of pol II transcripts. Class II genes that show wild-type patterns of expression in the absence of linker histones include the genes for TBP, histone H3 and the large subunit of pol II. Nevertheless, the knockout cells fail to downregulate a gene called *ngoA* under growth conditions in which it would normally be repressed. Another growth-responsive gene behaves normally under repressive conditions but is compromised in its ability to become activated in the H1-depleted cells. Thus, in *Tetrahymena* histone H1 does not perform a major global role in controlling transcription, but instead serves as a gene-specific regulator, with a stimulatory effect in one case and an inhibitory influence in another.

It is unclear to what extent the unexpected results obtained with *Tetrahymena* are truly representative of the general situation in eukaryotes. The histone H1 in these ciliates lacks the central globular domain found in H1 polypeptides of multicellular organisms. Nevertheless, its size, solubility and lysine richness is typical of the class, and it certainly plays an important role in chromatin condensation. Convincing evidence that histone H1 is also a gene-specific regulator in higher organisms comes from studies performed on pol III transcription in developing frog embryos. There are two families of active 5S rRNA genes in *Xenopus laevis*. One family is referred to as the somatic 5S genes, of which there are 400 copies per haploid genome, organized in a single cluster. The other is the oocyte 5S gene family, which has over 20 000 members. There are only six nucleotide differences between the 120-bp coding regions of these two types, but their flanking sequences are completely unrelated. Both sets of genes are transcribed at high levels during oogenesis, resulting in a massive accumulation of 5S rRNA for subsequent incorporation into ribosomes. (The genes encoding large rRNA are amplified specifically in oocytes, as a means to achieve the same end.) As already described, there is a general silencing of transcription during the first 12 cleavage divisions that follow fertilization. When transcription resumes at the mid-blastula transition, the oocyte 5S genes are expressed 50-fold less actively than the somatic 5S gene family. During the next two or three cell divisions, the oocyte 5S genes undergo further repression until their level of expression is reduced to 1000-fold less than the somatic 5S genes. This huge differential is then maintained for the rest of the frog's life. The mechanism of this striking regulation has been a subject of intensive study, as it is considered to provide a paradigm for situations in which promoters with similar but not identical *cis*-acting sequences recognized by the same group of factors can be controlled differentially. Because 5S rRNA genes have simple promoters and require relatively few transcription factors, it was hoped that they would provide a tractable system that would nevertheless reveal important insights into more complex regulatory situations. After considerable effort, this wish has been fulfilled. It has been found that the oocyte 5S genes are selectively subject to chromatin-mediated repression. This was demonstrated elegantly by manipulating the level of histone H1 *in vivo*.

When developing frogs were made to overexpress histone H1 by injecting its mRNA, the oocyte 5S genes became repressed prematurely. Conversely, when frogs were depleted of endogenous histone H1, by injecting a ribozyme that causes specific degradation of the H1 mRNA, repression of the oocyte 5S genes was substantially alleviated. These effects were highly specific; synthesis of somatic 5S rRNA and tRNA by pol III or of U1 and U2 snRNA by pol II did not respond to these manipulations of the linker histone pool (Fig. 8.5).

One reason why the oocyte 5S genes are especially sensitive to repression by histones is that they only assemble unstable transcription complexes. Once a preinitiation complex has formed on a tRNA or somatic 5S rRNA gene promoter, then the factors remain bound for a considerable period of time. Indeed, such complexes can remain intact for days or even weeks in nucleated *Xenopus* erythrocytes, which are transcriptionally inactive and devoid of RNA polymerase; this was shown by the ability of nuclei isolated from these cells to synthesize tRNA and 5S rRNA when supplemented with purified pol III. By contrast, the complexes that assemble on oocyte 5S rRNA genes are much less stable; the factors dissociate regularly and are in dynamic equilibrium with the factors in solution. In the absence of competition and when factors are plentiful, the oocyte 5S genes are transcribed as efficiently as the somatic type. However, in a developing frog they must compete with other genes for limiting factors; their inability to sequester a stable complex places them at a severe disadvantage in this regard. Furthermore, whenever the preinitiation complex dissociates from a promoter, histones are liable to gain access in its place. This problem is aggravated considerably by the fact that when a nucleosome assembles on an oocyte 5S rRNA gene promoter, it is positioned over the internal control region. Incorporation of histone H1 completely obscures the C-block element that is the primary recognition site for TFIIIA. The somatic 5S gene promoter also contains a strong nucleosome positioning signal, but in this case the C-block remains open so that TFIIIA can gain access even in the presence of histones; it is therefore much better equipped to resist the repressive encroachment of chromatin. As frogs develop after the MBT, the availability of TFIIIA diminishes. At the same time, a linker histone variant that is present during early development becomes replaced by the somatic form of histone H1, which has much higher affinity for nucleosomal DNA. These circumstances conspire to silence selectively the oocyte 5S rRNA genes.

Fig. 8.5 (*opposite*) Changes in the level of histone H1 in developing frog embryos causes selective regulation of oocyte 5S rRNA genes. Injection of mRNA encoding histone H1 causes an increase in the abundance of this linker histone and repression of oocyte-type 5S rRNA synthesis. Conversely, injection of a ribozyme that degrades the endogenous mRNA encoding histone H1 results in a decrease in the level of this linker histone and increased expression of the oocyte 5S rRNA genes. Transcription of genes encoding tRNA, somatic 5S rRNA, U1 or U2 snRNA is unaffected by these manipulations.

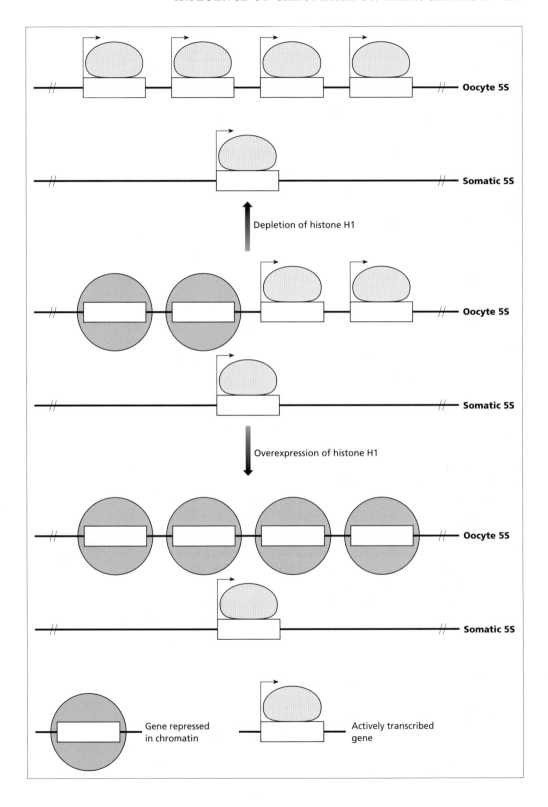

Oocyte 5S

Somatic 5S

Depletion of histone H1

Oocyte 5S

Somatic 5S

Overexpression of histone H1

Oocyte 5S

Somatic 5S

Gene repressed
in chromatin

Actively transcribed
gene

Activation of genes in chromatin

Under normal conditions, the *PHO5* gene of *S. cerevisiae* is packaged into an array of positioned nucleosomes that is only interrupted for ~80 bp within the promoter. *PHO5* encodes a secreted acid phosphatase that is used to scavenge extracellular phosphate when internal pools become exhausted. Phosphate starvation induces *PHO5* expression by at least 50-fold. Induction is accompanied by a profound rearrangement of four of the nucleosomes that were positioned at the promoter. This was revealed by the fact that ~600 bp of promoter DNA become hypersensitive to nuclease digestion. Treatment with DNase I or micrococcal nuclease is frequently used to investigate chromatin structure, because DNA that is incorporated into nucleosomes is significantly less susceptible to digestion, whereas open regions are preferentially cut and therefore appear to be 'hypersensitive' (Box 8.1). The chromatin disruption that accompanies induction of *PHO5* expression is thought to be a prerequisite of transcriptional activation, rather than a consequence, because mutation of the TATA box virtually abolishes transcription but does not prevent the transition in chromatin structure. Instead, the nucleosomal rearrangements are thought to be provoked by a **b**asic **h**elix-**l**oop-**h**elix (bHLH)-containing factor

Box 8.1 Micrococcal nuclease digestion of chromatin

Micrococcal **n**uclease (MNase) cleaves DNA in a sequence-independent fashion. If the DNA is assembled into chromatin, the regions in contact with histones are partially protected from digestion. MNase cleaves the most accessible regions preferentially; first the exposed linker DNA between the nucleosomes is digested, and only subsequently does it begin to attack the DNA within the nucleosome. Mild treatment with limited amounts of MNase can therefore be used to analyse the organization of chromatin and to isolate nucleosomes for further study. After partial digestion with MNase, naked DNA gives rise to a broad smear on agarose gels, reflecting the presence of fragments of many possible sizes. If chromatin is treated in the same way, a series of discrete bands will appear reflecting the length of DNA in a nucleosomal repeat. The smallest band will contain 146 bp, which is the length of DNA involved in the strongest contacts with the histone core. The diagram compares the appearance of naked DNA (lane 1) and chromatin (lane 2) samples after they have been treated with MNase and then resolved on an agarose gel. The DNA is visualized under UV light after staining with ethidium bromide. Alongside is illustrated the positions of preferential cutting (arrows) in a nucleosomal array that would give rise to the pattern shown in lane 2.

Box 8.1 (contd.)

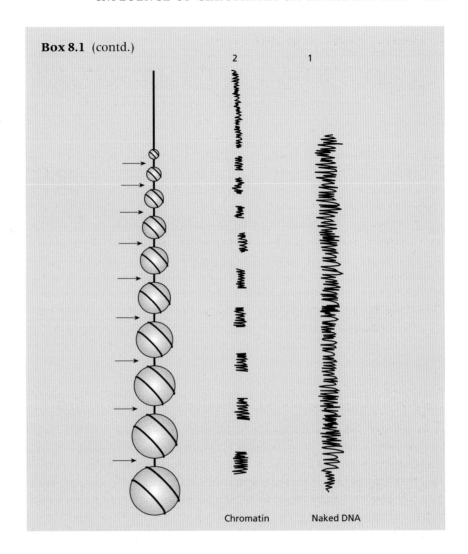

Chromatin   Naked DNA

called Pho4. Binding sites for Pho4 are found at −360 and −250, and these are critical for the activity of the *PHO5* promoter. *In vivo* footprinting reveals that Pho4 occupies these sites when phosphate is scarce, but not under high phosphate conditions when the promoter is silent. Indeed, Pho4 is found predominantly in the cytoplasm when phosphate is abundant (see Chapter 10). When the *PHO5* gene is repressed, the TATA box and the −250 Pho4-binding site are occupied by a pair of positioned nucleosomes that disappear following induction (Fig. 8.6). Displacement of these nucleosomes appears to be important for activation of the promoter. If the Pho4-binding sites are deleted, partial activation of TATA-directed transcription can be achieved by repressing histone synthesis. However, physiological levels of activation require the involvement of Pho4 and are not achieved by simply removing histones from the TATA box. When a DNA fragment with unusually high nucleosome-

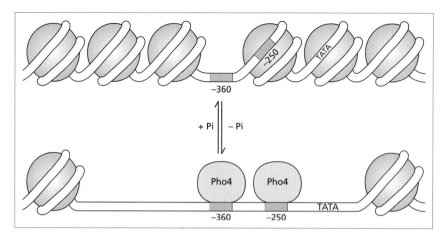

Fig. 8.6 Nucleosome positioning at the *PHO5* promoter. Under repressed conditions (+Pi), nucleosomes are positioned over most of the promoter, including the TATA box and the −250 binding site for Pho4. When phosphate is scarce, the *PHO5* gene is induced and four of the positioned nucleosomes are removed.

binding affinity was placed alongside the Pho4 recognition site, promoter inducibility was compromised severely. In the wild-type promoter, the Pho4 site at −360 remains free of nucleosomes even under high phosphate repressed conditions. It is thought that binding of Pho4 to this site is instrumental in rearranging the adjacent nucleosomes. Thus, mutation of the −360 site results in a promoter in which the chromatin structure persists even after phosphate starvation. When *PHO5* becomes activated, two nucleosomes on either side of the −360 Pho4 site are disrupted together in an 'all or nothing' manner. These four nucleosomes might therefore be regarded as a chromatin microdomain that is disrupted coordinately by Pho4.

Incorporation of the Pho4 recognition sequence into a nucleosome provides a potent barrier to factor binding. Access of Pho4 to the *PHO5* promoter therefore relies on the −360 site, which is maintained in a nucleosome-free state. However, some 'pioneer' factors, such as Gal4 and the **g**lucocorticoid **r**eceptor (GR), are able to reach their target DNA motifs even when they are assembled into a nucleosome. Indeed, if it is appropriately positioned, the presence of a nucleosome makes little difference to the affinity of the GR for its binding sites. This has been studied extensively in the context of the **m**ouse **m**ammary **t**umour **v**irus (MMTV). The MMTV promoter is induced by glucocorticoids and this requires GR binding, disruption of the local chromatin structure and assembly of an initiation complex around the TATA box. In the repressed state, the MMTV promoter is organized into an array of positioned nucleosomes. One of these incorporates the TATA box, whereas another incorporates two important recognition sites for a factor called NF1 (Fig. 8.7). Like Pho4, NF1 is

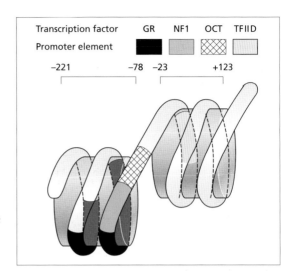

Fig. 8.7 Nucleosome positioning at the MMTV promoter. Recognition sites for TFIID, NF1 and the glucocorticoid receptor (GR) are incorporated into nucleosomes. TFIID and NF1 are unable to bind under these circumstances. However, the GR can gain access to its response elements and this disrupts the local chromatin structure, thereby freeing the TATA box and NF1 sites. Adapted from Wolffe 1995.

unable to recognize DNA that is in contact with histones. However, binding by the GR displaces histone H1, disrupts the nucleosome and allows NF1 access, which leads to the assembly of a preinitiation complex. The unusual ability of the GR to interact with glucocorticoid response elements that are incorporated into nucleosomes, despite the substantial distortion of DNA, is dependent on the recognition motifs being orientated away from the histone core (Fig. 8.4). The MMTV promoter contains four glucocorticoid response elements; two of these face towards the nucleosome core and are inaccessible to the GR.

The staged assembly of nucleosomes with respect to DNA sequence is a common feature of chromatin structure around promoters and enhancers. For factors such as Pho4, which cannot bind to their recognition sites if they are distorted by the presence of histones, positioning is important to keep binding motifs free of nucleosomes. In other cases, such as the GR, it is necessary to position the histones so that the DNA response element faces away from the core of the nucleosome. Under some circumstances, a well-positioned nucleosome can actually serve to stimulate transcription. An example is provided by the *Xenopus* vitellogenin B1 promoter, where reconstitution of a nucleosome can increase activity by approximately 10-fold. This effect is seen if the nucleosome is located in between an NF1-binding site at −120 and response elements for the oestrogen receptor at −300. It is thought that the nucleosome provides a scaffold which juxtaposes the oestrogen receptor with factors bound in proximity to the start site. (Fig. 8.8).

The DNA sequences that determine the positions of nucleosomes are not well understood, although there is a marked bias towards locations that place A/T-rich sequences in the minor groove next to the core region and G/C-rich sequences facing away from the core. Sequence-specific DNA-binding proteins can also contribute to the positioning of adjacent nucleosomes. In such

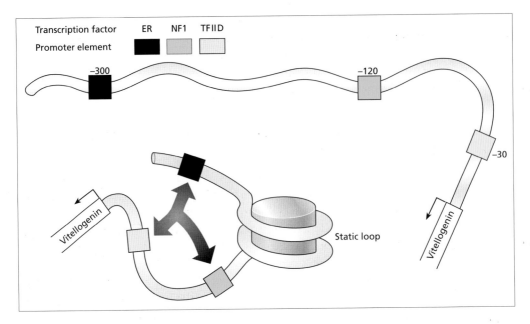

Fig. 8.8 A positioned nucleosome can stimulate transcription from the *Xenopus* vitellogenin B1 promoter. The core histones are thought to serve a scaffolding function which brings the oestrogen receptors (ER) into proximity with NF1 and the basal transcription factors. Adapted from Wolffe 1995.

cases, the continuous presence of the DNA-binding factor may be necessary to maintain nucleosome positioning. This was suggested by experiments which used the *E. coli* lac repressor protein bound to a pair of operator sites to dictate the location of five adjacent nucleosomes. When the lac repressor was induced to dissociate from DNA, the nucleosomes rearranged within 15 minutes. This striking mobility was dependent on ATP; in its absence, the same nucleosomes remained static for at least an hour after removal of the lac repressor.

Experiments carried out with truncated versions of Pho4 revealed that its activation domain is required for disruption of chromatin at the *PHO5* promoter. A Pho4 chimera containing the VP16 activation domain is also competent to reorganize chromatin. Likewise in the *GAL1* promoter, the disruption of a nucleosome adjacent to a cluster of Gal4-binding sites is dependent upon activation domains. These observations raised the possibility that chromatin rearrangement might simply occur in response to transcriptional activation. However, several lines of evidence show that this is not the case. For example, chromatin remodelling can occur in the absence of transcription at a crippled *PHO5* promoter with a mutated TATA box. In *Xenopus* microinjection experiments, a heterodimer of the thyroid hormone receptor (TR) and the 9-*cis* retinoic acid receptor (RXR) will disrupt nucleosomes over several hundred base pairs within the inducible TRβA promoter. TR mutants have been generated which retain the ability to reorganize chromatin in the absence of hormone but have lost the capacity to activate transcription upon hormone addition. Such mutants uncouple the processes of nucleosome remodelling and transcriptional activation and demonstrate clearly that promoter accessibility is insufficient to induce gene expression.

Transcription factors that resemble the structural proteins in chromatin

Because the proteins that package DNA into chromatin have coevolved with the transcription machinery, it is perhaps to be expected that certain components of these two groups will display structural and/or functional similarities. A striking example of this is provided by TFIID. The $hTAF_{II}31$ subunit of TFIID bears clear homology to histone H3, whereas $hTAF_{II}80$ resembles histone H4. In addition to the homologous regions, both TAFs have extended C-terminal tails that interact with other components of TFIID and also with transcriptional activators. Like histones H3 and H4 in the nucleosome, $hTAF_{II}31$ and $hTAF_{II}80$ exist as a heterotetramer in the TFIID complex. Furthermore, X-ray crystallography has shown that their *Drosophila* equivalents adopt a histone fold. In addition, $hTAF_{II}20$ and its *Drosophila* homologue show weak similarity with histone H2B; it can also bind to histones H3 and H4, as well as their TAF_{II} homologues. None of the TAFs have been found to resemble histone H2A, but each TFIID complex contains four molecules of $hTAF_{II}20$ and this polypeptide can homodimerize. It has therefore been proposed that TFIID may contain a nucleosome-like substructure, comprising $(hTAF_{II}20)_2–(hTAF_{II}31/hTAF_{II}80)_2$ $–(hTAF_{II}20)_2$ (Fig. 8.9). Protein–protein interaction analyses are consistent with this hypothesis. Disruption studies carried out both *in vitro* and *in vivo* have demonstrated that components of this presumptive TAF_{II} octamer are central to the architecture of TFIID. The DNase I footprinting pattern produced when TFIID binds to the adenovirus major late promoter resembles that generated by nucleosomal wrapping. These observations have lead to the proposal that $TAF_{II}s$ assemble core promoter DNA into a nucleosome-like structure. This might maintain the initiation region in a semicompacted state that is competent for transcription.

The ability of TFIID to adopt a nucleosomal architecture might confer substantial stability within the chromosomal environment. Outside of S phase, when the DNA is replicated, it seems very difficult to disrupt core histone interactions within nucleosomes *in vivo*. TFIID remains associated with chromosomes during mitosis, when transcription is silenced and chromatin becomes highly condensed. Indeed, the presence of TFIID has been inferred by genomic footprinting at a number of promoters in the absence of expression. An example in humans is provided by the gene for interleukin 2 (IL-2), which is only expressed in activated T-lymphocytes. *In vivo* footprinting showed that

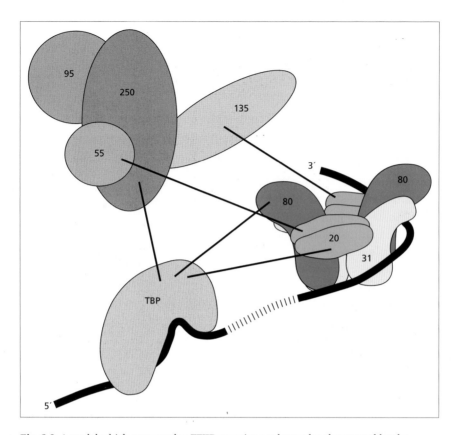

Fig. 8.9 A model which suggests that TFIID contains a subcomplex that resembles the nucleosome core, comprising a central tetramer of hTAF$_{II}$31/hTAF$_{II}$80 pairs that is flanked by (hTAF$_{II}$20)$_2$ dimers. This proposal is based on the fact that histones H2B, H3 and H4 bear homology to hTAF$_{II}$20, hTAF$_{II}$31 and hTAF$_{II}$80, respectively, in their histone fold regions. Furthermore, X-ray crystallography has shown that the *Drosophila* equivalents of hTAF$_{II}$31 and hTAF$_{II}$80 adopt a similar structure to histones H3 and H4. Protein–protein interaction studies are consistent with the putative nucleosome-like subcomplex, and suggest that it may be anchored to the remainder of TFIID by contacts with TBP, hTAF$_{II}$135 and hTAF$_{II}$55. It remains to be determined whether the TFIID subcomplex becomes wrapped by 146 bp of DNA, as is the case for the histone octamer. Adapted with permission from *Nature* **380**. Copyright Macmillan Magazines Limited.

IL-2 transcription correlates with the occupancy of upstream binding sites for activator proteins (e.g. NF-AT). By contrast, the TATA region is occupied in both resting and stimulated T cells. The footprinting pattern in the core promoter showed only minor changes when transcription was induced. Thus, the presence or absence of TFIID at the TATA box need not be diagnostic of ongoing gene expression. In some cases, it seems to indicate the existence of a poised state, in which promoters are inactive but rapidly inducible. However, TFIID is not detected at the IL-2 promoter in cells that are incapable of ex-

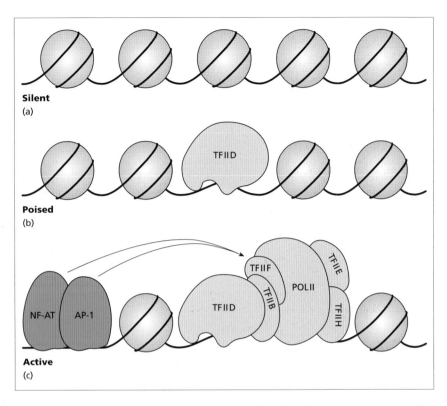

Fig. 8.10 States of the IL-2 promoter under distinct physiological circumstances. (a) In cells that are incapable of synthesizing IL-2 (e.g. HL-60 cells) TFIID is not detected at the IL-2 promoter. (b) In resting T-lymphocytes TFIID is bound to the IL-2 promoter, but the gene is not expressed; it may be considered to be in a 'poised' state. (c) Activation of T cells with ionomycin or phorbol esters triggers the binding of activators to upstream sites, recruitment of the remaining basal factors and initiation of IL-2 transcription.

pressing this gene (Fig. 8.10). Similarly, the permanent inactivation of the cell type-specific STE2 gene in yeast alpha cells correlates with the positioning of a nucleosome over the TATA region. Nucleosome formation at the core promoter is mutually exclusive of TFIID recruitment.

Several other transcription factors have been found to resemble histones. For example, the repressors Dr1 and DRAP1 display homology to histones H2A and H2B in the histone fold region and coexist as a heterodimer. It is striking that these cofactors regulate pol II transcription by forming a complex with TFIID. As already mentioned, the hepatic factor HNF3 contains a winged helix domain that is similar to those found in linker histones such as H1. In its active state, the mouse serum albumin enhancer exists in an array of precisely positioned nucleosome-like particles. HNF3 forms part of one of these particles and can direct their positioning with respect to DNA. It seems likely that HNF3 replaces linker histones within chromatin containing the serum albumin enhancer, thereby establishing a precise nucleoprotein architecture. In addition to positioning the nucleosomes, this may protect the enhancer from the repressive effects of histone H1. Because linker histones are retained less stably than core histones in nucleosomal DNA, the structure of HNF3 may be well adapted to a reversible regulatory role.

The yeast pol I activating protein UAF provides a particularly dramatic illustration of the fact that the eukaryotic transcription apparatus has evolved in parallel with the packaging components of chromatin. Purification of UAF revealed that two of its subunits are histones H3 and H4; however, histones H2A and H2B are not detected. Although the stoichiometry of H3 and H4 has not been determined in UAF, their presence may allow the factor to wrap DNA in a manner resembling a histone tetramer.

As described in Chapter 6, the pol I transcription system has changed relatively rapidly during evolution. UAF is not found in frogs or mammals. Instead, the factor UBF performs an architectural role in moulding the promoter region, as well as its function in recruiting SL1. The sequence of UBF is completely unrelated to that of UAF or any of the histones. However, the HMG boxes that are found in UBF are strongly homologous to domains present in the **h**igh **m**obility **g**roup (HMG) proteins, from which they take their name. Several HMG members are conserved and highly abundant constituents of metazoan chromatin. For example, there is approximately one molecule of HMG1 for every 20 nucleosomes in an average mammalian cell, and it is even more enriched in *Xenopus* eggs. HMG1 has been shown to bind to linker DNA and repress the transcription of 5S rRNA genes. The structural transitions that accompany interaction of HMG1 with a nucleosome are very similar to those produced by histone H1: HMG1 can constrain the path of the linker DNA and protect it from micrococcal nuclease; it can also position a nucleosome and restrict its mobility. Despite these functional similarities, HMG1 is unrelated to the linker histones in either sequence or structure. Apart from UBF, HMG domains are also found in a variety of other transcription factors, including LEF-1 and the sex-determining factor SRY.

These various examples demonstrate that regulatory transcription factors can resemble the structural components of chromatin in a number of ways. A potential advantage of this feature may be to allow transcription complexes to be maintained during the assembly of higher order chromosomal structures. Expanses of nucleosome-free DNA or large non-nucleosomal complexes might be expected to interfere with the process of chromatin folding. Perhaps the easiest way to allow transcription without disrupting chromosomal organization was for factors to evolve such that they can adopt nucleosome-like structures. An important feature of the histones is that they allow DNA to be exposed at the surface of nucleosomes. One side of the DNA is obstructed by the histones, but the other is exposed and potentially accessible to other regulatory proteins. Histone-like interactions between transcription factors and DNA may therefore facilitate the assembly of multicomponent complexes.

Chromatin remodelling complexes

The examples above provide ample evidence that changes in the level of transcription sometimes demand the reorganization of chromatin associated with particular genes. This may be especially true for the core region of pol II promoters. TBP can bind to TATA sequences in naked DNA with a Kd of 10^{-9}–10^{-10} M, but is unable to bind at all to certain chromatin templates at concentrations as high as 10^{-6} M. The elongation phase of transcription can also be significantly impaired by the presence of nucleosomes. A number of chromatin remodelling complexes have been identified, which may help to alleviate such problems.

The prototype chromatin remodelling factor is a multisubunit complex of ~2 MDa called SWI/SNF. Components of this complex were originally isolated from *S. cerevisiae* in genetic screens to identify factors involved in mating type switching (SWI is short for switch). Mutations in SWI/SNF genes can disrupt the transcription of a subset of yeast genes, including the *HO* gene which encodes an endonuclease that is required to switch mating types. Large SWI/SNF complexes have also been found in *Drosophila* and humans, and these contain homologues of many of the yeast subunits. Indeed, truncations of human SNF5 have been linked to malignant rhabdoid tumours, an extremely aggressive type of paediatric cancer. The effects of mutations in yeast SWI/SNF genes can often be suppressed by secondary mutations in histones H3 and H4 or HMG-1. Reducing the level of histones H2A and H2B in cells can also partially suppress *swi/snf* phenotypes. These genetic effects lead to the idea that the SWI/SNF complex might regulate gene expression by counteracting chromatin-mediated repression. Support for this possibility came from studies of the yeast *SUC2* promoter *in vivo*. The *SUC2* gene encodes invertase, which is required for growth on sucrose. When it is induced, its upstream promoter becomes more sensitive to micrococcal nuclease, implying an opening of the chromatin in this region. Certain mutations in SWI/SNF genes prevent these chromatin changes and impair transcriptional induction; both effects can be alleviated by lowering the levels of histones H2A and H2B. The SWI/SNF-dependent reorganization of chromatin at the *SUC2* promoter still occurs if the TATA box is crippled, which shows that it is not simply a consequence of transcriptional induction.

Purification revealed that the SWI/SNF complex from yeast contains 11 subunits. One of these, SWI2/SNF2, has a DNA-dependent ATPase activity. The purified complex was found to facilitate binding of Gal4 derivatives or TATA-binding protein (TBP) to nucleosomal DNA in an ATP-dependent manner. It can also assist in the local disruption of preassembled nucleosomes, as revealed by changes in the DNase I digestion pattern. Similar effects have been observed with the human SWI/SNF complex, which has in addition been shown to help transcriptional activators to stimulate paused pol II in reading through downstream nucleosomes. SWI/SNF appears to remodel chromatin by catalysing the sliding or tracking of nucleosomes along DNA. There is also evidence that it can displace nucleosomes from a DNA template. DNA microarray analysis revealed that inactivation of SWI/SNF alters the expression levels of 6% of all genes in *S. cerevisiae*. In many cases, expression increased; although this could be an indirect effect, it raises the possibility that in some circumstances remodelling can result in repression instead of activation.

As described in Chapter 4, a proportion of the pol II found in cells is associated with many components of the basal transcription machinery in a large complex referred to as the holoenzyme. Cofractionation and coimmunoprecipitation experiments have provided evidence that the SWI/SNF complex also interacts with the pol II holoenzyme. However, these data are controversial, because SWI/SNF can also be isolated free from the holoenzyme components; it is unclear whether this is due to dissociation during isolation or because only some of the SWI/SNF associates with pol II in cells. One study calculated that there are ~1000–2000 copies of the SWI/SNF complex per yeast cell, which is similar to the abundance of the holoenzyme. However, another estimate suggested that SWI/SNF is 10-fold less abundant. If SWI/SNF does bind to the holoenzyme *in vivo*, then they may be recruited together by transcriptional activators. Such a mechanism has obvious appeal, as it might allow the opening of chromatin and the recruitment of the basal machinery to occur in a concerted fashion. The model needs to explain why SWI/SNF mutations have gene-specific effects if the complex is associated with the basal machinery. A possible explanation is that some promoters, such as *PHO5*, recruit the holoenzyme and displace nucleosomes so robustly that SWI/SNF is not required; other promoters, such as *SUC2*, might need a remodelling factor because the chromatin structure is more refractory or recruitment is less efficient. Alternatively, different promoters may employ distinct methods for disrupting nucleosomes.

A number of additional chromatin remodelling complexes have been identified. Like SWI/SNF, these are all large multisubunit factors. They can disrupt nucleosomes in an ATP-dependent manner and some have been shown to facilitate transcription. For example, RSC (which stands for **r**emodelling the **s**tructure of **c**hromatin) is a 15-subunit complex that was isolated from *S. cerevisiae* on the basis of homology with several SWI/SNF components. Unlike SWI/SNF, RSC is encoded by essential genes and is not found associated with the pol II holoenzyme. It is also moderately abundant, with 10^3–10^4 copies per cell, perhaps suggesting that it performs a less specialized function than SWI/SNF.

A factor called NURF (which stands for **nu**cleosome **r**emodelling **f**actor) was purified from extracts of *Drosophila* embryos because of its ability to facilitate chromatin disruption by transcription factors. The 500 kDa NURF complex consists of four polypeptides. One of these displays significant similarity to the ATPase domain of SWI2/SNF2; as a consequence, it was named ISWI, for **i**mitation **swi**tch. Homologues of ISWI have been identified in yeast and humans; the human and *Drosophila* versions are 75% identical over their entire length, a level of conservation which suggests an important function. NURF is a very efficient remodelling machine; a single NURF complex is sufficient to disrupt 18 nucleosomes. With preassembled chromatin templates, NURF facilitates transcriptional induction by Gal4 derivatives, as long as they contain an activation domain. Once it has performed its nucleosome remodelling function, NURF is no longer required for recruitment of the basal machinery and transcription of at least 100 bp. Additional chromatin remodelling complexes have also been isolated from *Drosophila* extracts; these are called ACF, for **A**TP-utilizing **c**hromatin assembly and remodelling **f**actor, and CHRAC, for **ch**romatin **r**emodelling and **a**ssembly **c**omplex. Although distinct from NURF, these complexes also contain ISWI and can mobilize nucleosomal arrays in an ATP-dependent manner. Other remodelling factors have also been identified, including the FACT and NURD complexes that will be described subsequently. Indeed, databases of genomic DNA sequences show that there are still many uncharacterized proteins with strong homology to the ISWI and SWI2/SNF2 ATPases. It is therefore quite likely that further chromatin remodelling factors are waiting to be discovered.

One of the human proteins with homology to SWI2/SNF2 is called ATRX. It is the product of a gene in which mutations give rise to a severe form of syndromal mental retardation characterized by the presence of α-thalassaemia, urogenital abnormalities and facial dysmorphism (ATR–X syndrome). Inadequate synthesis of α-globin is responsible for the thalassaemic phenotype. It is therefore thought that ATRX regulates a discrete set of targets, which includes the α-globin genes.

As well as obscuring the promoter region, nucleosomes can also provide a very effective block to the passage of RNA polymerases during transcript elongation. A complex has been purified from human cells that helps pol II to overcome this roadblock. It is composed of two polypeptides, of 140 kDa and 80 kDa, and is referred to by the acronym FACT, which stands for **fa**cilitates **c**hromatin **t**ranscription. FACT allows efficient passage of pol II through nucleosomes during transcript elongation, but does not function on a naked DNA template. It does not require ATP hydrolysis and operates in a similar stoichiometry to the nucleosomes. FACT interacts specifically with histone H2A/H2B dimers and appears to promote nucleosome disassembly during transcription. Indeed, FACT activity can be blocked by covalently crosslinking nucleosomal histones. The 80-kDa subunit of FACT contains an HMG domain, which may allow it to bind DNA where it enters and exits the nucleosome. Given the tremendous number of nucleosomes that need to be passed during transcription of a large gene, FACT or complexes with similar functions can be expected to be very important in allowing expression *in vivo*. It is estimated that there are more than 10^5 molecules of FACT in a HeLa cell.

Acetylation

Specific lysines in the tails of histones are often found to be acetylated in chromosomal regions containing genes that are active or potentially active, such as the β-globin locus in erythroid cells. For example, core histone hyperacetylation and general DNase I sensitivity is associated with a 33-kb region encompassing the chicken β-globin locus. Conversely, silenced regions such as heterochromatin or the mating-type loci in *S. cerevisiae* contain histones that are underacetylated. The chromatin immunoprecipitation, or ChIP, assay has often been used to investigate whether nucleosomes are acetylated at a particular chromosomal site; this technique is described in Box 8.2. Chemicals that increase histone acetylation are potent inducers of HIV-1 transcription in latently infected T cells. Acetylation appears to function at several levels to influence gene expression. Higher-order folding of chromatin can be disrupted. Furthermore, binding of transcription factors to nucleosomal DNA can be significantly facilitated if the histones are acetylated. One simplistic idea is that the acetyl groups neutralize the positive charge of lysines within histone tails and thereby weaken the electrostatic interactions with the negatively charged phosphodiester backbone of DNA. This cannot be the whole story, because substitution of lysine with either basic arginine or neutral glutamine residues can result in similar phenotypes. It is probable that acetylation disrupts secondary structure within the histone tails. Furthermore, there is evidence that acetylation can alter the conformation of nucleosomes and can also diminish interactions between adjacent nucleosomes, thereby favouring

Box 8.2 The chromatin immunoprecipitation (ChIP) assay

This is a modification of the coimmunoprecipitation assay that was introduced in Box 3.1. It is used to investigate whether a protein X, such as hyperacetylated histone H3, is bound to a particular region of chromosomal DNA, such as a promoter element. Cells are treated with formaldehyde, which crosslinks histones to DNA so that they are not released during the subsequent manipulations. The cells are then sonicated to shear the DNA into fragments of ~200–1000 bp in length. The crosslinked and sonicated chromatin is incubated with an anti-protein X antibody that has been immobilized on insoluble agarose beads. After the beads have been pelleted by centrifugation, they are washed and then the immunoprecipitated material is eluted. **P**olymerase **c**hain **r**eactions (PCRs) are subsequently carried out using primers that are specific to the DNA region of interest. If the PCR detects that DNA region in the immunoprecipitate, this would suggest that it is bound by protein X in chromatin. For example, an antibody against hyperacetylated histones is likely to immunoprecipitate promoter DNA from highly active genes but not DNA from silent regions of chromosomes.

Box 8.2 (contd.)

Chromatin immunoprecipitation (ChIP)

To ask if protein X binds to DNA region Y in chromatin:

1 Cells are treated with formaldehyde to crosslink proteins to DNA

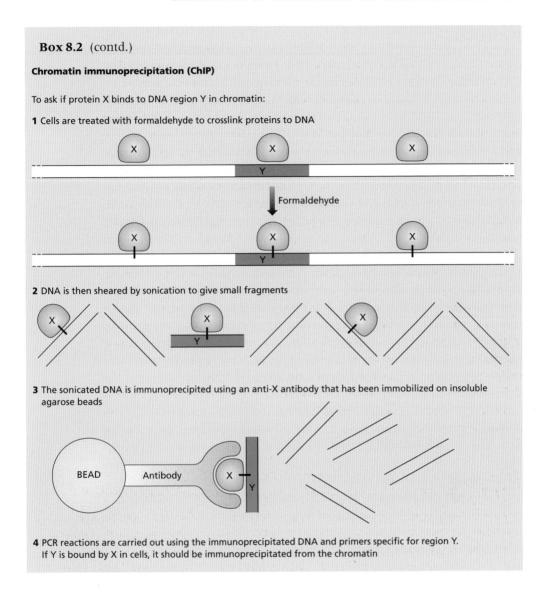

Formaldehyde

2 DNA is then sheared by sonication to give small fragments

3 The sonicated DNA is immunoprecipited using an anti-X antibody that has been immobilized on insoluble agarose beads

BEAD   Antibody

4 PCR reactions are carried out using the immunoprecipitated DNA and primers specific for region Y. If Y is bound by X in cells, it should be immunoprecipitated from the chromatin

decompaction. It is also possible that the acetyl groups serve as signals for proteins that are capable of disrupting chromatin.

The yeast protein GCN5 was identified in genetic screens as a transcriptional coactivator. Recombinant GCN5 was found to display **h**istone **a**cetyl**t**ransferase (HAT) activity, acetylating histone H3 strongly and histone H4 weakly. However, isolated GCN5 can only acetylate free histones. In yeast it associates with proteins called Ada2 and Ada3 to form complexes that are capable of acetylating nucleosomes. The largest of these complexes is called SAGA—**S**pt-**A**da-**G**cn5-**a**cetyltransferase. In addition to GCN5 and the Ada proteins, SAGA contains several products of *spt* genes that were isolated as suppressors of transcription defects caused by promoter insertions of the transposable element Ty. Several of the TAF (TBP-associated factor) components of TFIID are also found within SAGA, even though the complex does not contain TBP. The HAT catalytic domain of GCN5 is required for its coactivator function *in vivo*. Mutations in GCN5, Ada2 or Ada3 slow growth and reduce the ability of acidic activators such as VP16 to stimulate transcription. Indeed, Ada2 binds to the activation domains of VP16 and Gal4. TBP also interacts with Ada2 or Ada3 in yeast extracts. These observations have lead to a model in which transcriptional activation in a chromatin context may involve the initial binding of an activator such as Gal4, which then recruits a complex with HAT activity; this acetylates nucleosomes, thereby increasing the accessibility of promoter DNA, and also serves a bridging function which helps recruit the basal machinery (Fig. 8.11).

A more complex picture emerged from *in vivo* studies of the *HO* gene, which encodes an endonuclease that is required for mating type switching in *S. cerevisiae*. Genetic analysis showed that activation of the *HO* promoter requires both GCN5 and the SWI/SNF remodelling complex, as well as two transcription factors called SWI5 and SBF. The order in which these proteins operate was deduced by immunoprecipitating chromatin at various times during the induction process. The first factor to bind is SWI5, a zinc finger protein which can penetrate the closed chromatin environment and find two DNA sites located ~1.2 kb upstream of the *HO* gene. SWI5 triggers the recruitment of SWI/SNF, which is detected almost immediately in the vicinity of the SWI5-binding sites. About 5 minutes later, SWI/SNF can also be found in proximity to a series of SBF recognition sites, which are located between 700 bp and 100 bp upstream of the transcription start site. However, SBF binding requires the additional involvement of GCN5, which is recruited after SWI/SNF. Thus, whereas SWI5 can penetrate chromatin unaided, SBF needs the assistance of both HAT and remodelling complexes. Recruitment of the basal transcription machinery and expression of the *HO* gene only occurs once SBF is in position. A particularly interesting feature of this process is that SWI5 dissociates from the *HO* promoter almost immediately after it has brought in the SWI/SNF complex (Fig. 8.12). The concerted action of a HAT and a remodelling complex that occurs in this example is likely to be a recurring theme. Because acetylation is a covalent modification, it may serve to fix a remodelled state after the action of complexes such as SWI/SNF.

The HAT GCN5 is well conserved through evolution and has been identified in humans. It is also related to the metazoan coactivator P/CAF, which can acetylate both free and nucleosomal histones. As was described in Chapter 5, P/CAF binds to the coactivators CBP and p300, which can acetylate all four core histones either free or as nucleosomes. CBP and p300 are found in multicellular organisms as diverse as man and worms, but do not occur in yeast. In addition, the $TAF_{II}250$ subunit of TFIID has been shown to acetylate histones; this enzymatic activity might be important for accessing the TATA

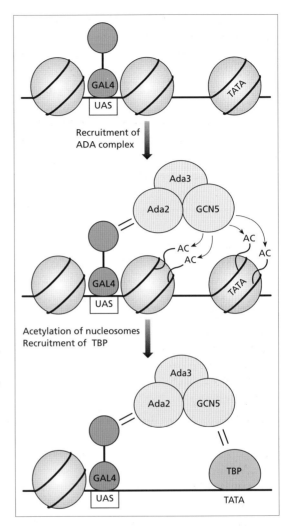

Fig. 8.11 Model for the function of the Ada coactivator complex in *S. cerevisiae*. An activator such as Gal4 interacts with Ada2 through its acidic activation domain, thereby recruiting the complex. The histone acetyltransferase (HAT) activity of GCN5 transfers acetyl groups (Ac) to the N-terminal tails of histones, thereby reducing the ability of nucleosomes to occlude the promoter region. TBP can then gain access to the TATA box, possibly aided by interactions with the Ada complex. GCN5, Ada2 and Ada3 can all be found as part of the much larger complex SAGA.

box region, or it may help maintain the accessibility of the promoter region once TFIID has been recruited. However, like GCN5, TAF$_{II}$250 acetylates free histones H3 and H4 rather than nucleosomes. Two other mammalian HATs are the nuclear receptor coactivators ACTR and SRC-1; because these can associate with P/CAF, which in turn binds to CBP and p300, it would seem that coactivator complexes can form involving multiple HATs.

Acetylation is believed to play an important role in transcriptional regulation by the thyroid hormone and retinoic acid receptors (TR and RAR, respectively). In the presence of the appropriate ligand, these receptors recruit coactivator complexes that include CBP, P/CAF and other HATs, which increase the accessibility of the surrounding chromatin. However, TR and RAR in the absence of ligand play an active and dominant role in silencing

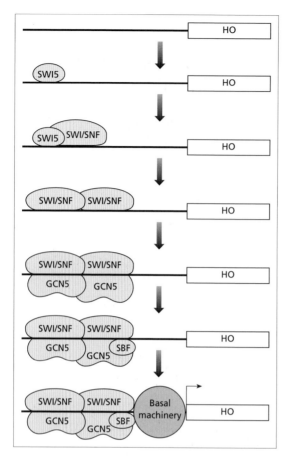

Fig. 8.12 The order of events leading to transcriptional activation of the *HO* gene in *S. cerevisiae*. SWI5 initiates the induction process by binding to its recognition sites despite the presence of a 'closed' chromatin environment. SWI5 recruits the SWI/SNF remodelling complex and then rapidly dissociates from the promoter. After 5–10 minutes, SWI/SNF can also be detected further downstream. GCN5 arrives soon after SWI/SNF; it is likely to be recruited as part of the SAGA complex. The transcription factor SBF gains access to its DNA sites only after SWI/SNF and GCN5-containing complexes have opened the chromatin at the *HO* promoter. Finally, SBF helps recruit the basal transcription machinery, allowing expression of the *HO* gene.

transcription; this serves to amplify the magnitude of response to ligand by eliminating any 'leaky' expression that might otherwise occur. Promoter silencing by these receptors involves corepressor proteins, an example of which is N-CoR (**n**uclear **r**eceptor **co**repressor). N-CoR binds to ligand-free receptor in the flexible hinge region which connects the DNA-binding and ligand-binding domains; mutations in this region that prevent N-CoR binding also abolish silencing. N-CoR is found in a complex with two other cofactors called Sin3 and RPD3; blocking experiments with microinjected antibodies have shown that all three components of this complex are necessary for the silencing effect. RPD3 is a **h**istone **deac**etylase (HDAC) and is therefore thought to diminish chromatin accessibility by removing acetyl groups from the histone tails, thereby discouraging the recruitment of the basal transcription machinery. However, N-CoR also displays some repressive activity that is independent of its ability to recruit Sin3 and RPD3. Binding of ligand results in conformational changes that trigger the release of N-CoR and the recruitment of coactivators with HAT activities (Fig. 8.13). Thus, TR and RAR can act to make chromatin either opaque or transparent to other transcription factors by

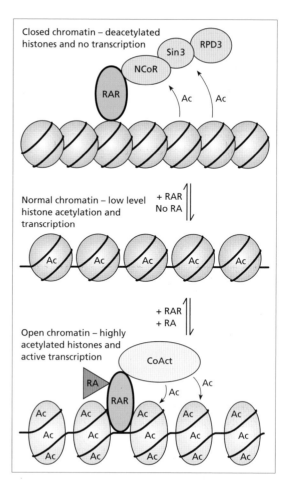

Fig. 8.13 The retinoic acid receptor (RAR) can function as either an activator or a silencer of transcription, depending on whether or not its ligand retinoic acid (RA) is available. In the absence of RA, the RAR is bound by the corepressor N-CoR or a related corepressor called SMRT (**s**ilencing **m**ediator for **R**XR and **T**R). These in turn interact with the repressor proteins Sin3 and RPD3. RPD3 has histone deacetylase activity and is thought to use this to render the chromatin region opaque to other transcription factors. Binding of RA induces conformational changes in the RAR which release N-CoR along with Sin3 and RPD3 and instead recruit a coactivator complex. Several HATs can be found in the coactivator complex, including CBP, p300, P/CAF, ACTR and SRC-1; together these are thought to increase chromatin accessibility by acetylating histones, thereby facilitating transcription complex assembly.

alternatively recruiting HATs or an HDAC in a ligand-dependent manner. In addition, studies in microinjected frog oocytes have shown that TR has supplementary effects upon chromatin structure that are independent of the addition or removal of acetyl groups.

In acute promyelocytic leukaemias, chromosomal translocations some-times result in the RARα receptor becoming fused to either the **pro**myelocytic **l**eukaemia (PML) or **p**romyelocytic **l**eukaemia **z**inc **f**inger (PLZF) proteins. The resultant fusion proteins retain the DNA-binding and ligand-binding domains of RARα and can regulate RAR target genes, but appear to impair the differentiation of promyelocytic cells. Treatment with high doses of retinoic acid can induce the differentiation of leukaemic blast cells carrying the PML–RARα fusion, and this results in disease remission. However, the PLZF–RARα-containing leukaemias are resistant to retinoic acid. Both fusion proteins recruit the N-CoR/Sin3/RPD3 histone deacetylase complex. Whereas the complex is released from PML-RARα by retinoic acid, the PLZF–RARα fusion retains the complex in the presence of ligand due to a ligand-independent

N-CoR-binding site in the PLZF part of the fusion (Fig. 8.14). A chemical inhibitor called trichostatin A, which specifically blocks deacetylase activity, converts PLZF–RARα from an inhibitor to an activator of retinoic acid-induced differentiation. This suggests that deacetylation is crucial to the oncogenic function of the fusion protein, presumably due to the repression of genes that are necessary for the switch from proliferation to differentiation. These observations have raised the possibility that deacetylase inhibitors might be used therapeutically to treat patients with acute promyelocytic leukaemias, especially those who do not respond to retinoic acid. It is striking that another form of acute myeloid leukaemia is associated with a chromosomal translocation that fuses CBP to a putative HAT called MOZ (**mo**nocytic-leukaemia **z**inc finger). It therefore seems that deregulated acetylation is a recurrent theme during leukaemogenesis.

The bHLH–Zip proteins Mad and Max form heterodimers that repress transcription in competition with heterodimers composed of Max and the cellular oncoprotein Myc. Mad acts in part by competing with Myc for the shared dimerization partner Max. However, the Mad–Max complex also serves as a potent transcriptional repressor by recruiting Sin3 and the associated histone deacetylase RPD3. Both Sin3 and RPD3 are necessary for Mad to antagonize the transforming activity of Myc. A third example of a transcription factor that can utilize histone deacetylase activity in order to repress transcription is the tumour suppressor protein RB, the product of the retinoblastoma susceptibility gene. Unlike Mad and N-CoR, RB can bind to HDAC1 in the absence of Sin3. RB represses genes encoding thymidine kinase and dihydrofolate reductase, enzymes which are necessary for DNA synthesis and cell proliferation. The histone deacetylase inhibitor trichostatin A blocks the ability of RB to repress these genes. Although these results clearly implicate deacetylation as a mechanism that is used by RB in certain cases, other genes can be repressed by RB irrespective of the presence of trichostatin A.

Despite these important examples, the overwhelming majority of genes appear not to be responsive to histone acetylation. A differential display analysis revealed that trichostatin A altered the expression of only eight genes out of 340 that were monitored. Similarly, in *S. cerevisiae* the deletion of deacetylase Rpd3, which is related to HDAC1, increases the overall level of core histone acetylation but only alters the expression of a small subset of genes. It may be that changes in acetylation need to work alongside other forms of regulation in order to influence the transcriptional state of most genes. The

Fig. 8.14 (*opposite*) Chromosomal translocations in acute promyelocytic leukaemias generate RAR fusion proteins that can be defective in their response to retinoic acid (RA). A fusion protein containing the promyelocytic leukaemia protein (PML) responds to RA by releasing the N-CoR histone deacetylase complex and recruiting coactivators with histone acetyltransferase activity. However, an RAR fusion protein containing the promyelocytic zinc finger protein (PLZF) retains N-CoR even in the presence of RA and continues to silence transcription despite the recruitment of coactivators.

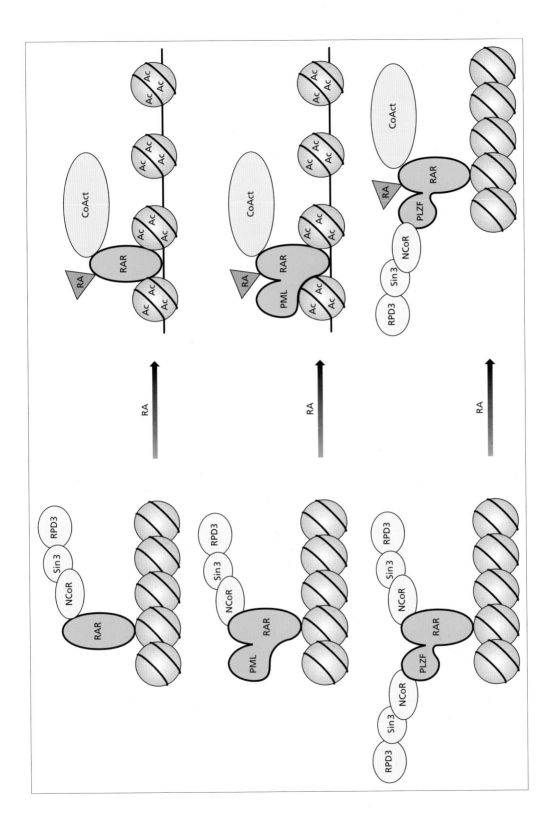

fact that relatively few promoters respond to changes in acetylation is consistent with the substantial evidence that changes in ubiquitous chromatin components can produce regulatory effects that are highly gene specific.

Contrary to expectations, Rpd3 deletion was found to reduce transcription near telomeres in yeast and have a similar effect near heterochromatin in *Drosophila*. The general correlation between histone acetylation and transcriptional activity is therefore clearly an oversimplification of what is certainly a complex method of control. This has been well illustrated in *Drosophila*, where the lysines of histone H4 that become acetylated can vary according to the chromosomal region. Whereas H4 acetylated at positions 5 or 8 is distributed widely throughout euchromatin, acetylation of H4 at residue 16 is associated preferentially with the hyperactive male X chromosome. Furthermore, lysine 12 of histone H4 is acetylated preferentially in the repressed heterochromatin of both *Drosophila* and *Saccharomyces*. Other proteins besides histones can also be targets for acetyltransferases. For example, P/CAF, p300 and $TAF_{II}250$ can all acetylate the β subunit of TFIIE. The DNA-binding activity of p53 is stimulated by acetylation. It is therefore probable that some of the effects of HATs and deacetylases are histone independent. There is clearly still much to learn concerning the involvement of acetylation in transcriptional control.

Methylation

In the adult cells of vertebrates, 60–90% of CpG dinucleotides are methylated on the cytosine. High levels of methyl-CpG are generally found at genomic regions which are transcriptionally silent and packaged into nuclease-resistant chromatin. Methylation is believed to repress transcription by at least two different mechanisms. For certain factors, such as CREB, the presence of a methyl group can interfere with binding to a DNA recognition site. However, patches of DNA containing a high density of methyl-CpG can confer repression upon adjacent unmethylated promoters. In such cases, the inhibitory effects are thought to be mediated by proteins that recognize the methylated cytosine. Under certain conditions, histone H1 can show preferential binding to methylated DNA. Furthermore, transcriptionally inactive chromatin that is rich in methyl-CpG also contains elevated levels of histone H1. Nevertheless, much of the regulatory effect of methyl-CpG is thought to result from dedicated DNA-binding proteins that recognize this sequence specifically. The best characterized example of such a factor is an abundant mammalian protein called MeCP2. This factor contains a DNA-binding domain with specificity for methylated CpG and a distinct repressor domain that can function independently when fused to Gal4. The repressor domain has been shown to interact with Sin3 and thereby recruit HDAC1 onto promoters; this is expected to inhibit transcription by deacetylating histones and reducing accessibility (Fig. 8.15). Indeed, the deacetylase inhibitor trichostatin A diminishes the ability of MeCP2 to repress gene expression, although some residual inhibitory effect remains.

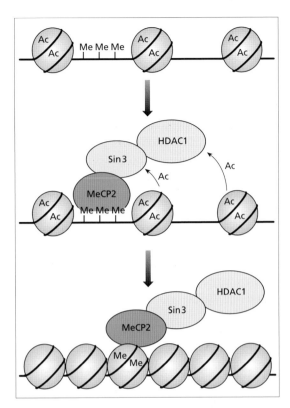

Fig. 8.15 Transcriptional repression by MeCP2 involves deacetylation. Regions of DNA containing methylated CpG sequences (Me) are bound preferentially by MeCP2. MeCP2 recruits Sin3 and HDAC1 to these regions and the latter removes acetyl groups (Ac) from histones; this is expected to reduce promoter accessibility and perhaps promote chromatin compaction.

Gal4–VP16 can normally penetrate chromatinized templates, even in the presence of histone H1, but this is not the case if the chromatin is assembled on methylated DNA. It is therefore thought that methylation provides a means to lock chromatin securely in a silenced state.

De novo mutations in MeCP2 have been found to be responsible for some cases of Rett syndrome, a progressive neurodevelopmental disorder that is one of the most common causes of mental retardation in females. In several cases the syndrome has resulted from missense mutations in the methyl-CpG-binding domain. Two other cases have involved a frameshift and a nonsense mutation which disrupt the transcription repression domain. Such mutations may be lethal in males, who would not be able to compensate with a wild-type copy of the gene for MeCP2 because it is carried on the X-chromosome.

At least four other mammalian proteins have a **m**ethyl-CpG-**b**inding **d**omain (MBD) that resembles the one found in MeCP2. One of these proteins, MBD3, resides in a macromolecular entity called the **nu**cleosome **r**emodelling histone **d**eacetylase complex or NURD. NURD contains eight subunits altogether, another of which is Mi-2, an autoantigen associated with the human connective tissue disease dermatomyositis. Mi-2 contains a motif homologous to the DNA helicase/ATPase domain of SWI2/SNF2. A third component of the complex is HDAC1; indeed, NURD is the most abundant deacetylase complex in mammalian cells. The complex displays remodelling activity by virtue of Mi-2, and this promotes deacetylation by helping HDAC1 gain access to nucleosomal histones. Thus, in contrast to the examples described previously, remodelling in this case is associated with transition to a repressed chromatin state. It is believed that NURD interacts with methylated DNA by means of its MBD3 subunit, but it can also be brought to other regions of chromatin through protein–protein interactions with sequence-specific DNA-binding factors.

Methylated CpG sequences are found on the inactive X chromosome of female placental mammals. However, at least 90% of the methyl-CpG in a mammalian genome occurs within silent transposable elements and retroviruses. It has been suggested that DNA methylation evolved as a defence mechanism to provide protection against genomic parasites such as these. An example may be provided by the Alu family of repetitive class III genes, which is thought to have spread by retrotransposition. Alu genes are heavily methylated in a wide range of somatic human cell types. Indeed, it has been estimated that chromatinization silences ~99% of potentially active Alu promoters. This may be extremely important, because the 500 000 Alu copies in the haploid human genome constitutes 5% of all chromosomal DNA. This enormous number of templates could provide a massive sink for transcription factors, which might compete with essential genes encoding tRNA and 5S rRNA and deprive them of shared factors, such as TFIIIB and TFIIIC. This problem is circumvented by assembling most Alu promoters into inaccessible regions of chromatin.

Because methyl groups persist at specific sites during DNA replication, methylation provides an ideal mechanism for ensuring that a gene remains repressed during passage through the cell cycle; this is not the case for nucleosomes, which are displaced by the DNA replication fork. In higher organisms such as vertebrates, which have large numbers of tissue-specific genes, DNA methylation may also provide a mechanism for switching off permanently the genes that are not necessary in a particular differentiated cell type. Such stable silencing of a large fraction of the genome would allow the transcription machinery to be focused where it is most required. Undifferentiated mouse embryonic stem cells can survive without MeCP2 or any detectable DNA methylation; however, DNA methyltransferase is required once the cells begin to differentiate. In mouse embryos, MeCP2 and normal DNA methylation levels are essential for postgastrulation development, whereas blastocysts divide in the absence of methylated DNA. Clearly, other mechanisms of gene regulation can compensate for the lack of methylation in these undifferentiated cell types. The same is true for *S. cerevisiae* and *D. melanogaster*, which have no detectable methyl-CpG in their genomes. DNA methylation must have evolved relatively recently, presumably to provide a supplementary level of gene control in complex organisms.

Fragile-X syndrome is a frequent cause of mental retardation, affecting about one in 6000 males. It almost always results from the expansion of a triplet CGG repeat in the 5' untranslated region of the fragile-X mental retardation (*FMR1*) gene. Expansion beyond ~200 repeats generally coincides with hypermethylation of the region containing the *FMR1* promoter. The severity of mental retardation correlates with the degree of methylation, rather than the length of CGG expansion. Chromatin changes in the expanded, methylated region suggest a closed, transcriptionally silent state, consistent with the lack of *FMR1* transcripts in severely affected individuals. Patients with a mild phenotype usually exhibit mosaicism in both the degree of methylation and the level of *FMR1* expression. Congenital myotonic dystrophy has also been linked with increased methylation and altered chromatin structure following the expansion of a triplet repeat.

Aberrant methylation is thought to be an important step during the formation of certain types of tumour. In order for cancers to develop, at least one tumour suppressor gene must become inactivated. Very frequently this occurs through genetic mutation. However, in some cases it is found that the neoplastic cells have silenced wild-type tumour suppressor genes through *de novo* methylation of their promoter regions. For example, the *Rb* gene is repressed due to hypermethylation in a small fraction of retinoblastomas and the *VHL* gene is aberrantly methylated in a subset of sporadic renal carcinomas. A far more common target for this phenomenon is the *p16* tumour suppressor gene, which encodes a small polypeptide that can arrest proliferation by binding and inactivating the cyclin D-dependent kinases which help drive the cell cycle. One study found that the promoter region of p16 had undergone *de novo* methylation in ~20% of primary neoplasms. This was especially prevalent in certain common tumour types, such as carcinomas of the lung, head and neck. It seems likely that this mechanism for silencing key protective genes contributes to a large number of cancer cases.

Chromosomal domains

Within chromosomes, the compacted chromatin fibres are organized into large domains. The structural basis of this organization is poorly understood. There does not seem to be a simple correlation between the edges of a chromosomal domain, as determined by structure, and the locations of genes. Several genes are often found inside a domain, but in some cases a domain boundary appears to lie within a gene. Immunofluorescence microscopy has revealed regional variations in the content of linker histones and the degree of core histone acetylation. Certain regions of chromosomes stain darkly when examined cytologically and are highly condensed throughout the cell cycle. Such domains are referred to as heterochromatin; they are transcriptionally inert, generally late replicating, underacetylated and enriched in methylcytosine. Most genes within heterochromatin are maintained in a stably silent state, although they can escape repression. Heterochromatin is found at centromeres and telomeres and also, in *S. cerevisiae*, at the silent mating loci. An extreme example of heterochromatin is the inactive X chromosome in female mammals, where gene expression is suppressed in order to compensate for double dosage. In *Drosophila*, heterochromatin occupies >30% of the genome. Within it, histone H4 is acetylated preferentially at lysine 12, which has led to the suggestion that heterochromatin formation may be regulated by specific acetylation events.

The repressive effects of heterochromatin can sometimes spread into adjacent regions of normal chromatin and silence nearby genes. The extent of this spreading can vary from one cell to another, giving rise to a phenomenon called 'position effect variegation'. This was first discovered in *Drosophila*, when chromosomal rearrangements placed the *white*⁺ eye colour gene in the vicinity of heterochromatin; some flies were found to have eyes with patches of red adjacent to patches of white, indicating that expression of *white*⁺ varied from cell to cell. That variegation is a chromatin effect is suggested by the fact that its extent can be modified by changing the dosage of histone genes or by altering the level of histone acetylation either chemically or genetically.

Certain sequences called insulators can prevent the spreading of heterochromatin. The first example of such structures was provided in *Drosophila* by specialized chromatin structures (scs) that are found at the junctions between the open chromatin of the transcriptionally active 87A7 heat-shock locus and the adjacent condensed chromatin. Each scs is defined by a pair of nuclease hypersensitive sites flanking a 250–300-bp DNA segment. Scs elements independently have neither a positive nor a negative effect upon transcription. However, they can establish a domain of independent gene activity which is insulated from neighbouring regions by the scs present at each border. An scs element can also block communication between an enhancer and a promoter. Thus, they can insulate loci against either the repressive effects of heterochromatin or the stimulatory effects of enhancers (Fig. 8.16). The molecular mechanism of this phenomenon is not well understood.

Position effects have constituted a major problem for research involving transgenic mice, because the random integration of microinjected DNA can often place reporter genes at sites where their transcription is partially or totally repressed. As a consequence, expression levels can vary dramatically

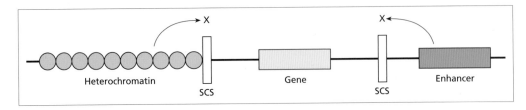

Fig. 8.16 A genetic locus that is flanked by scs elements is insulated against the effects of neighbouring chromosomal regions, including the spreading of heterochromatin or activation by enhancers.

between different mice carrying the same integrated reporter. Studies of the human β-globin locus led to the discovery of a locus control region (LCR), which confers physiological levels of erythroid-specific expression on linked transgenes irrespective of their site of integration. The ability of LCRs to override position effects suggests that they can open silent chromatin and establish and maintain an environment that is competent for transcription. A deletion that removes the β-globin LCR is responsible for many cases of thalassaemia in hispanic populations. Chromosomal domains as large as 100 kb can be activated by LCRs.

The LCR that controls human β-globin expression has been characterized extensively. It is located over 50 kb upstream of the globin genes and appears as a series of five erythroid-specific DNase I hypersensitive sites within a contiguous piece of DNA. Each of these discrete regions contains a core of 150–300 bp with a high density of binding motifs for transcription factors, some of which are themselves erythroid specific. The mechanism of action of LCRs is far from clear and somewhat controversial. However, the five individual hypersensitive regions of the β-globin LCR appear to act together as a functional unit that interacts with a single gene in the locus at any particular moment by looping out the intervening DNA. Initiation from a competent promoter within the open chromatin domain seems to be triggered by direct contact with the LCR, and continued association is needed to maintain transcription.

Further reading

Reviews

Armstrong, J. A. & Emerson, B. M. (1998) Transcription of chromatin: these are complex times. *Curr. Opin. Genet. Dev.* **8**: 165–172.

Berger, S. L. (1999) Gene activation by histone and factor acetyltransferases. *Curr. Opin. Cell Biol.* **11**: 336–341.

Bird, A. P. & Wolffe, A. P. (1999) Methylation-induced repression-belts, braces, and chromatin. *Cell* **99**: 451–454.

Cairns, B. R. (1998) Chromatin remodeling machines: similar motors, ulterior motives. *Trends Biochem. Sci.* **23**: 20–25.

Fraser, P. & Grosveld, F. (1998) Locus control regions, chromatin activation and transcription. *Curr. Opin. Cell Biol.* **10**: 361–365.

Geiduschek, E. P. (1998) Chromatin transcription: clearing the gridlock. *Curr. Biol.* **8**: R373–R375.

Grant, P. A., Sterner, D. E., Duggan, L. J., Workman, J. L. & Berger, S. L. (1998) The SAGA unfolds: transcription regulators in chromatin-modifying complexes. *Trends Cell Biol.* **8**: 193–197.

Hayes, J. J. & Wolffe, A. P. (1992) The interaction of transcription factors with nucleosomal DNA. *Bioessays* **14**: 597–603.

Higgs, D. R. (1998) Do LCRs open chromatin domains? *Cell* **95**: 299–302.

Imhof, A. & Wolffe, A. P. (1998) Gene control by targeted histone acetylation. *Curr. Biol.* **8**: R422–R424.

John, S. & Workman, J. L. (1998) Just the facts of chromatin transcription. *Science* **282**: 1836–1837.

Kass, S. U., Pruss, D. & Wolffe, A. P. (1997) How does DNA methylation repress transcription? *Trends Genet.* **13**: 444–449.

Kingston, R. E. & Narlikar, G. J. (1999) ATP-dependent remodeling and acetylation as regulators of chromatin fluidity. *Genes Dev.* **13**: 2339–2352.

Knoepfler, P. S. & Eisenman, R. N. (1999) Sin meets NuRD and other tails of repression. *Cell* **99**: 447–450.

Kornberg, R. D. & Lorch, Y. (1999) Twenty-five years of the nucleosome, fundamental particle of the eukaryote chromosome. *Cell* **98**: 285–294.

Pazin, M. J. & Kadonaga, J. T. (1997) What's up and down with histone deacetylation and transcription? *Cell* **89**: 325–328.

Razin, A. (1998) CpG methylation, chromatin structure and gene silencing—a three-way connection. *EMBO J.* **17**: 4905–4908.

Struhl, K. (1998) Histone acetylation and transcriptional regulatory mechanisms. *Genes Dev.* **12**: 599–606.

Struhl, K. (1999) Fundamentally different logic of gene regulation in eukaryotes and prokaryotes. *Cell* **98**: 1–4.

Sun, F.-L. & Elgin, S. C. R. (1999) Putting boundaries on silence. *Cell* **99**: 459–462.

Svaren, J. & Horz, W. (1997) Transcription factors vs. nucleosomes: regulation of the *PHO5* promoter in yeast. *Trends Biochem. Sci.* **22**: 93–97.

Tyler, J. K. & Kadonaga, J. T. (1999) The 'dark side' of chromatin remodeling: repressive effects on transcription. *Cell* **99**: 443–446.

Wakimoto, B. T. (1998) Beyond the nucleosome: epigenetic aspects of position-effect variegation in *Drosophila*. *Cell* **93**: 321–324.

Wolffe, A. P. (1998) *Chromatin: Structure and Function*, 3rd edn. Academic Press, San Diego, CA.

Wolffe, A. P. & Pruss, D. (1996) Deviant nucleosomes: the functional specialization of chromatin. *Trends Genet.* **12**: 58–62.

Selected papers

Structure of the nucleosome

Luger, K., Mader, A. W., Richmond, R. K., Sargent, D. F. & Richmond, T. J. (1997) Crystal structure of the nucleosome core particle at 2.8 Å resolution. *Nature* **389**: 251–260.

Gene regulation by histones

Bouvet, P., Dimitrov, S. & Wolffe, A. P. (1994) Specific regulation of *Xenopus* chromosomal 5S rRNA gene transcription *in vivo* by histone H1. *Genes Dev.* **8**: 1147–1159.

Han, M. & Grunstein, M. (1988) Nucleosome loss activates yeast downstream promoters *in vivo*. *Cell* **55**: 1137–1145.

Panetta, G., Buttinelli, M., Flaus, A., Richmond, T. J. & Rhodes, D. (1998) Differential nucleosome positioning on *Xenopus* oocyte and somatic 5S RNA genes determines both TFIIIA and H1 binding: a mechanism for selective H1 repression. *J. Mol. Biol.* **282**: 683–697.

Prioleau, M.-N., Huet, J., Sentenac, A. & Mechali, M. (1994) Competition between chromatin and transcription complex assembly regulates gene expression during early development. *Cell* **77**: 439–449.

Schild, C., Claret, F.-X., Wahli, W. & Wolffe, A. P. (1993) A nucleosome-dependent static loop potentiates oestrogen-regulated transcription from the *Xenopus* vitellogenin B1 promoter *in vitro*. *EMBO J.* **12**: 423–433.

Shen, X. & Gorovsky, M. A. (1996) Linker histone H1 regulates specific gene expression but not global transcription *in vivo*. *Cell* **86**: 475–483.

Chromatin remodelling machines

Kwon, H., Imbalzano, A. N., Khavari, P. A., Kingston, R. E. & Green, M. R. (1994) Nucleosome disruption and enhancement of activator binding by a human SWI/SNF complex. *Nature* **370**: 477–481.

Mizuguchi, G., Tsukiyama, T., Wisniewski, J. & Wu, C. (1997) Role of nucleosome remodeling factor NURF in transcriptional activation of chromatin. *Mol. Cell* **1**: 141–150.

Orphanides, G., LeRoy, G., Chang, C.-H., Luse, D. S. & Reinberg, D. (1998) FACT, a factor that facilitates transcript elongation through nucleosomes. *Cell* **92**: 105–116.

Orphanides, G., Wu, W.-H., Lane, W. S., Hampsey, M. & Reinberg, D. (1999) The chromatin-specific transcription elongation factor FACT comprises human SPT16 and SSRP1 proteins. *Nature* **400**: 284–288.

Owen-Hughes, T., Utley, R. T., Cote, J., Peterson, C. L. & Workman, J. L. (1996) Persistent site-specific remodeling of a nucleosome array by transient action of the SWI/SNF complex. *Science* **273**: 513–516.

Whitehouse, I., Flaus, A., Cairns, B. R., White, M. F., Workman, J. L. & Owen-Hughes, T. (1999) Nucleosome mobilization catalysed by the yeast SWI/SNF complex. *Nature* **400**: 784–786.

Wilson, C. J., Chao, D. M., Imbalzano, A. N., Schnitzler, G. R., Kingston, R. E. & Young, R. A. (1996) RNA polymerase II holoenzyme contains SWI/SNF regulators involved in chromatin remodeling. *Cell* **84**: 235–244.

Activation of the HO promoter

Pia Cosma, M., Tanaka, T. & Nasmyth, K. (1999) Ordered recruitment of transcription and chromatin remodeling factors to a cell cycle-and developmentally regulated promoter. *Cell* **97**: 299–311.

Acetylation

Grignani, F., De Matteis, S., Nervi, C. *et al.* (1998) Fusion proteins of the retinoic acid receptor-α recruit histone deacetylase in promyelocytic leukaemia. *Nature* **391**: 815–818.

Kuo, M.-H., Zhou, J., Jambeck, P., Churchill, M. E. A. & Allis, C. D. (1998) Histone acetyltransferase activity of yeast Gcn5p is required for the activation of target genes *in vivo*. *Genes Dev.* **12**: 627–639.

Lee, D. Y., Hayes, J. J., Pruss, D. & Wolffe, A. P. (1993) A positive role for histone acetylation in transcription factor access to nucleosomal DNA. *Cell* **72**: 73–84.

Luo, R. X., Postigo, A. A. & Dean, D. C. (1998) Rb interacts with histone deacetylase to repress transcription. Cell **92**: 463–473.

Turner, B. M., Birley, A. J. & Lavender, J. (1992) Histone H4 isoforms acetylated at specific lysine residues define individual chromosomes and chromatin domains in *Drosophila* polytene nuclei. *Cell* **69**: 375–384.

Wong, J., Patterton, D., Imhof, A. *et al.* (1998) Distinct requirements for chromatin assembly in transcriptional repression by thyroid hormone receptor and histone deacetylase. *EMBO J.* **17**: 520–534.

Methylation

Jones, P. L., Veenstra, G. J. C., Wade, P. A. *et al.* (1998) Methylated DNA and MeCP2 recruit histone deacetylase to repress transcription. *Nature Genet.* **19**: 187–191.

Nan, X. *et al.* (1998) Transcriptional repression by the methyl-CpG-binding protein MeCP2 involves a histone deacetylase complex. *Nature* **393**: 386–389.

Stoger, R., Kajimura, T. M., Brown, W. T. & Laird, C. D. (1997) Epigenetic variation illustrated by DNA methylation patterns of the fragile-X gene *FMR1*. *Hum. Mol. Genet.* **6**: 1791–1801.

Wade, P. A., Gegonne, A., Jones, P. L. *et al.* (1999) Mi-2 complex couples DNA methylation to chromatin remodelling and histone deacetylation. *Nature Genet.* **23**: 62–66.

Locus control regions

Wijgerde, M., Grosveld, F. & Fraser, P. (1995) Transcription complex stability and chromatin dynamics *in vivo. Nature* **377**: 209–213.

Chapter 9: Controlling Transcription Factor Production

In order to control the expression of its genes, a cell needs to regulate its content of transcription factors. This process is clearly fundamental to determining its specific phenotypic properties. Many of the factors encoded by its DNA will not be required in a particular cell type. Changing the constellation of factors present in a cell will often lead to an altered phenotype, as occurs, for example, when cells differentiate. Like every protein, the production of a transcription factor can be regulated at any of the steps leading from the gene to its product, including transcription, RNA processing and translation. People who work on transcription factors have a tendency to concentrate on transcriptional control, as this is generally their primary research interest, but post-transcriptional regulation can be extremely important and is likely to be much more prevalent than is currently realized. This chapter will provide examples of these various levels of control. Once a factor is synthesized, its specific activity can be regulated through another tier of mechanisms, but discussion of these will be deferred until Chapters 10 and 11.

Control of transcription of genes encoding transcription factors

As is the case for all gene products, transcription constitutes the primary level at which transcription factor synthesis is controlled. This provides the raw material for secondary levels of regulation, such as splicing and translation. Each of the principles that have already been described for the transcriptional control of class II genes can be expected to apply to the genes that encode transcription factors themselves. This will, of course, involve the combinatorial action of a range of proteins at a series of specific promoter and enhancer sites. There is therefore a regulatory hierarchy, in which one transcription factor determines the synthesis of another. In addition, it is common for factors to participate in the control of the gene(s) that encode them, thereby establishing a feedback sequence. Feedback control will be discussed in more detail later.

A good example of a hierarchy of transcription factors is seen in the liver and pancreas. In these tissues, cell type-specific gene expression is controlled primarily at the level of transcription and involves several factors, including C/EBPs (CCAAT/enhancer binding proteins) and various HNFs (hepatocyte nuclear factors). HNF-1α is present at much higher concentrations in liver and pancreas than it is in most other tissues and it stimulates the production of a number of hepatocyte gene products, such as albumin (Fig. 9.1). The gene encoding HNF-1α is, in turn, controlled by HNF-4α. HNF-1α and HNF-4α are

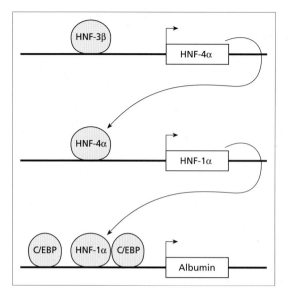

Fig. 9.1 The hierarchy of HNF transcription factors that operates in the liver and pancreas. The albumin gene promoter is activated by HNF-1α and C/EBPs that bind to sites close to the TATA box. The gene encoding HNF-1α is activated by HNF-4α. The promoter of the HNF-4α gene contains binding sites for HNF-3β, which is required for its activation. HNF-3 proteins have also been shown to modify the nucleosomal organization of the albumin enhancer.

important transcription factors clinically, because mutations in their genes impair insulin secretion and cause two phenotypically indistinguishable forms of type 2 diabetes, called MODY1 and MODY3. The MODY diabetic phenotype is thought to result from defects in the expression of genes involved in glucose metabolism. The promoter for HNF-4α contains evolutionarily conserved binding sites for HNF-3β, a factor with a winged helix DNA-binding domain that closely resembles linker histone H5. Targetted disruption of the gene encoding HNF-3β results in a substantial reduction in the synthesis of HNF-4α, HNF-1α and many downstream target gene products, such as albumin and various enzymes that participate in glycolysis and gluconeogenesis. HNF-3β is therefore believed to function as a positive regulator at the top of the HNF transcription factor hierarchy. This is consistent with the fact that it is expressed at the earliest stages of fetal development, well before HNF-4α. Treatment of cells with insulin can upregulate HNF-3β, as well as HNF-4α and many of the genes further down the hierarchy.

Another very good example of a hierarchy of transcription factors that control the expression of genes encoding other transcription factors occurs during *Drosophila* development, where Bicoid activates gap genes to produce transcription factors that regulate pair-rule genes, some of which themselves encode factors that regulate the homeotic genes which code for homeodomain-containing transcription factors. This complex cascade will be described in detail in Chapter 14.

Transcription factor dilution

An unusual mechanism for transcriptional regulation is found during the early period of *Xenopus* development. *Xenopus* oocytes stock-pile vast quantities of products, which sustain them through the initial stages of embryogenesis when biosynthesis is curtailed. For example, an oocyte accumulates suffi-cient core histones to assemble 13 000–16 000 genomes into chromatin. The ribosomal components are also produced at extremely high levels and this is achieved by a variety of mechanisms. Rapid synthesis of 18S, 5.8S and 28S rRNAs is possible because the class I genes encoding these transcripts are amplified specifically in oocytes. By contrast, all *Xenopus* cells contain a large excess of 5S rRNA genes, but most of these are only transcribed actively during oogenesis. As described in the previous chapter, the somatic 5S rRNA gene family consists of 400 copies per haploid genome, organized in a single cluster. In addition, each haploid genome contains 20 000 copies of the major oocyte and 1300 copies of the trace oocyte 5S rRNA gene. In order to ensure high levels of expression of these genes, an oocyte produces vast quantities of TFIIIA, the factor responsible for nucleating transcription complex assembly on 5S rRNA genes. Indeed, each *Xenopus* egg contains an astonishing 1.5 billion TFIIIA molecules, which works out as 70 000 molecules per 5S rRNA gene. This statistic becomes even more striking in light of the fact that, unlike most transcription factors, TFIIIA has a single genetic target. However, as was described in Chapter 7, TFIIIA also binds to the transcripts of the gene it regulates, and so much of the TFIIIA in oocytes is located in the cytoplasm in a complex with 5S rRNA.

Following fertilization, *Xenopus* embryos develop rapidly in the absence of any transcription. The total amount of TFIIIA in the embryo remains relatively constant, but as cell numbers increase and DNA is replicated, the number of TFIIIA molecules per genome falls dramatically. By the gastrula stage, there is only one TFIIIA molecule per 5S rRNA gene and this value continues to decline as development proceeds, despite the eventual synthesis of more factor. Eventually a steady state is reached in which TFIIIA is five-fold less abundant than the genes it regulates, a situation which persists in adult cells. Under these circumstances the oocyte 5S rRNA gene families are unable to compete for the limiting amounts of factor and they become incorporated into silent chromatin, in a process that is dependent on histone H1, as was described in Chapter 8. By contrast, the somatic 5S rRNA genes are able to recruit the avail-able TFIIIA and resist the repressive influence of histones. In effect, one of the major determinants of 5S rRNA synthesis during *Xenopus* development is the availability of a transcription factor that undergoes serial dilution.

Control of transcription factor production through alternative splicing

The primary transcripts of many class II genes contain non-coding intron sequences that must be removed by splicing to create a functional mRNA. In some cases the transcripts can be spliced in several different ways in order to generate multiple products from the same gene. This may give rise to mRNAs that differ in their protein coding sequences or transcripts with changes in their untranslated regions that can influence stability, translation efficiency or intracellular localization. Alternative splicing is often constitutive, producing the same mix of isoforms under all circumstances. However, there are also many examples in which the splicing pattern is regulated in order to change the genetic output according to cell type or conditions. This can be achieved through inhibitors that prevent the use of particular splice sites or activators that direct the splicing to alternative sites.

An example of a transcription factor that is subject to alternative splicing is WT1, the product of the Wilms' tumour suppressor gene. Wilms' tumour is a paediatric kidney cancer that affects 1 in 10 000 infants. In 10–15% of cases, Wilms' tumour patients display mutations in the *WT1* predisposition gene. Several acute leukaemias have been found to involve small insertions in *WT1* that are predicted to give rise to truncated polypeptide products. Mutations in *WT1* also cause nearly all cases of Denys–Drash syndrome, a rare developmental disorder of the urogenital system. In addition, *WT1* mutations cause Frasier's syndrome, an uncommon condition defined by male pseudohermaphroditism and progressive glomerulopathy. All of the Frasier-associated mutations in *WT1* arise in introns, an observation which might be explained by functional differences between the alternative splice variants.

Analysis of patients with *WT1* mutations and also of knockout mice has shown that WT1 is required for the normal development of the kidney and gonads. It can bind and regulate the promoters of a number of genes that are involved in controlling growth. The protein has four TFIIIA-type Cys_2His_2 zinc fingers at its C-terminus. The *WT1* gene consists of 10 exons that give rise to four protein isoforms through alternative splicing: exon 5 encodes 17 residues that can be either included or left out, and a three-residue lysine-threonine-serine (KTS) sequence may or may not be included between the third and fourth zinc fingers (Fig. 9.2). The KTS insertion disrupts the spacing between the last two fingers and reduces the affinity of WT1 for specific DNA sequences. Despite this, the isoforms that include the KTS sequence are expressed at higher levels than those that do not. Evidence of their physiological importance is provided by the fact that species as diverse as humans, frogs and zebrafish all produce both +KTS and –KTS isoforms. Frasier's syndrome patients have intronic mutations that reduce the +KTS/–KTS ratio; this suggests that production of the +KTS isoforms is important for urogenital development. Immunofluorescence and immunoprecipitation analyses have provided evidence that WT1 associates with both the transcriptional and splicing machineries in living mammalian cells. Furthermore, WT1 can bind to certain RNA sequences and this activity is influenced by the KTS sequence. The +KTS isoforms appear to interact preferentially with spliceosomes, whereas the –KTS isoforms are associated primarily with the transcriptional apparatus. Differential splicing may therefore help determine whether WT1 functions as a regulator of transcription or of RNA processing.

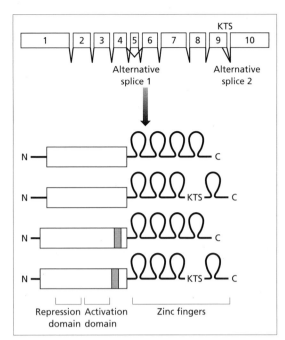

Fig. 9.2 The alternative splicing patterns of the *WT1* transcript and the four protein isoforms that can result. The sequences encoded by the 10 exons of *WT1* are spliced together in four different ways to generate distinct mRNAs, as indicated at the top. Exon 5 sequences are sometimes excluded to remove 17 amino acids from the regulatory domain of the WT1 protein. Alternative splicing can also determine whether or not the tripeptide KTS encoded by the end of exon 9 is included between the third and fourth zinc fingers of the WT1 polypeptide.

As described in Chapter 3, Pit-1 is a POU domain transcription factor that governs the ontogeny of the somatotroph, lactotroph and thyrotroph lineages of pituitary cells. It is required for the pituitary-specific expression of the prolactin, growth hormone and thyrotropin genes. An alternatively spliced form of Pit-1, called Pit-1β, has its activation domain disrupted by a 26-residue insertion. As a consequence, Pit-1β functions as a dominant negative inhibitor of Pit-1 function.

The related factors **c**AMP-**r**esponse **e**lement-**b**inding protein (CREB) and CREM are also encoded by multiple transcripts that are generated by alternative splicing. Both these proteins bind to **c**AMP-**r**esponsive **e**lements (CREs) in many hormonally regulated promoters. They also contribute to several physiological systems, such as memory, circadian rhythms and spermatogenesis. At least seven different splice variants of the CREB mRNAs have been detected. Some of these include novel exons with stop codons in all three reading frames and therefore produce C-terminally truncated CREB isoforms that lack the bZip domain. These polypeptides have no known function: they cannot bind DNA, are not localized to the nucleus and have no detectable effect upon cAMP-dependent transcription. They may constitute irrelevant by-products of differential splicing. Alternative splicing also gives rise to at least seven different isoforms of CREM (Fig. 9.3). Some of these function as transcriptional activators, whereas others serve as repressors owing to the exclusion of two glutamine-rich activation domains. The repressors can antagonize cAMP-induced transcription, either by binding to CREs as non-stimulatory

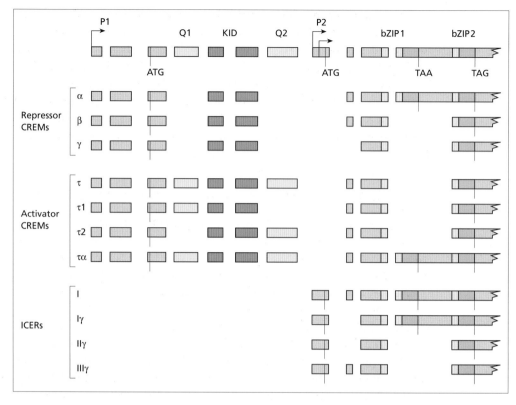

Fig. 9.3 Multiple isoforms of CREM are produced from a single gene by differential splicing. The various exons of the gene are shown above; these are spread over more than 80 kb of genomic DNA. Beneath are shown spliced transcripts that have been identified. Transcripts encoding the CREM proteins are all derived from the cAMP-independent P1 promoter. All these mRNAs encode a **k**inase-**i**nducible **d**omain (KID) that includes the cAMP-dependent phosphoacceptor site. Several encode one or two glutamine-rich regions and give rise to activating CREM protein isoforms. Others exclude these activation domains and generate transcriptional repressors. Two alternative bZIP DNA-binding domains can be used. A cAMP-inducible promoter P2 drives transcription initiation from an internal site 82 bp upstream of the eighth exon; this introduces an ATG translation start site and adds eight novel amino acids to the sequence encoded by exon 8; the short and alternatively spliced mRNAs derived from the P2 promoter encode a series of truncated polypeptides called the ICERs (**i**nducible **c**AMP **e**arly **r**epressors). The positions of the translation initiation (ATG) and alternative termination (TAA and TAG) sites are shown. Reproduced with permission from Paolo Sassone-Corsi.

homodimers or by heterodimerizing with the activators and thereby weakening their ability to increase expression. The *CREM* gene also encodes two separate bZip DNA-binding domains, either of which can be incorporated into its polypeptide products. Extensive differential splicing therefore allows this gene to single-handedly encode an entire family of factors with distinct regulatory properties.

The pattern of expression of CREM isoforms is developmentally regulated. In prepubertal testis CREM is present at very low levels and only the repressor forms are detected. However, mature germ cells of the adult testis express high levels of the CREMτ activator mRNA and protein. CREMτ is involved in stimulating the transcription of testis-specific haploid-expressed genes. Potential targets include genes that encode the transition proteins and protamines that replace histones in order to compact chromatin in spermatozoa. In knockout mice in which the CREM gene is disrupted, there are no mature germ cells or spermatogenesis.

Degradation of mRNAs encoding transcription factors

The half-life of a typical mRNA is 10–20 minutes in yeast and several hours in mammals. However, the turnover times of mRNAs can vary drastically, from a few minutes to days. Rapid degradation of mRNA provides an effective mechanism for ensuring that a protein is only expressed transiently. It is often achieved by incorporating into the transcript a destabilizing sequence such as AUUUA. Several copies of this sequence are found in the 3' untranslated regions of mRNAs encoding c-Myc, c-Jun and c-Fos. As a consequence, these transcription factors are synthesized only transiently following serum stimulation of quiescent cells. AUUUA elements destabilize transcripts in a copy-dependent manner by triggering their deadenylation; the half-life of β-globin mRNA, which is normally very stable, can be substantially reduced by artificially introducing these sequences.

CREM transcripts may contain up to 10 copies of AUUUA in their 3' untranslated regions, but this number can vary due to the use of alternative polyadenylation sites. During spermatogenesis, the most 5' polyadenylation site is utilized preferentially in response to follicle-stimulating hormone; this results in especially stable mRNAs due to exclusion of nine of the AUUUA elements. This can account for the very high levels of CREMτ transcript that accumulate in mature testes.

Translational control of transcription factor synthesis

Several important examples have been discovered in which transcription factor synthesis is regulated at the level of translation. Control of translation provides an additional regulatory checkpoint and offers the opportunity of a rapid response that bypasses the nucleus, relying on *cis*-regulatory sequences within the mRNA. The best known example of a transcription factor that is regulated translationally is GCN4, a bZip protein from *Saccharomyces cerevisiae* that activates at least 40 genes involved in the biosynthesis of amino acids and purines.

Eukaryotic translation initiation factor 2 (eIF2) is responsible for delivering charged initiator tRNA^{Met} to the 40S ribosomal subunit in a ternary complex with GTP. The resultant 43S preinitiation complex binds near the 5' cap of mRNA and migrates downstream until it reaches an AUG codon, where it is joined by the 60S ribosomal subunit allowing translation to commence. Starvation conditions trigger phosphorylation of the α subunit of eIF2, which results in a generalized translational inhibition. However, translation of the GCN4 mRNA bucks the trend and is stimulated, allowing induction of the biosynthetic pathways needed for the production of amino acids and purines. This differential response of GCN4 is because its mRNA contains four short **u**pstream **o**pen **r**eading **f**rames (uORFs) located 150–360 nucleotides 5' to the authentic initiation codon. When nutrients are plentiful, the uORFs inhibit GCN4 translation by restricting the progression of ribosomes that are scanning through the leader in search of the start codon. Eliminating the AUGs of the four uORFs results in constitutive high level expression of GCN4. The fourth uORF from the 5' end is sufficient on its own to reduce translation to only 1% of the level seen in the absence of all uORFs; this is because all the ribosomes translate uORF4 and then dissociate from the mRNA. By contrast, the first uORF alone decreases GCN4 translation by only 50%; this is thought to be because half the ribosomes are able to resume scanning and find the authentic AUG after translating uORF1. When all four uORFs are present and amino acids are abundant, all the ribosomes that resume scanning after uORF1 translation reinitiate at the downstream uORFs and then dissociate from the mRNA without synthesizing GCN4. However, under starvation conditions many ribosomes that resume scanning after translating uORF1 ignore the remaining uORFs and reinitiate at the GCN4 start site instead. Thus, prior translation of uORF1 allows ribosomes to overcome the strong translational barrier at uORF4. To explain these observations, it is proposed that under plentiful conditions ribosomes that resume scanning after uORF1 quickly rebind the eIF2.GTP.Met.tRNA^{Met} ternary complex before reaching uORF4, where they are obliged to reinitiate. But the concentration of these complexes is much lower under starvation conditions due to the phosphorylation of eIF2α. Many ribosomes therefore scan through the downstream uORFs without rebinding tRNA^{Met} and so cannot recognize their AUGs. However, many of these ribosomes will have rebound eIF2.GTP.Met.tRNA^{Met} by the time they reach the more distant GCN4 AUG, allowing them to reinitiate and synthesize the transcription factor. Thus, reducing the concentration of eIF2.GTP.Met.tRNA^{Met} complexes is thought to allow ribosomes to bypass the inhibitory uORFs 2–4 and find the GCN4 start site instead. This mechanism only allows GCN4 production under starvation conditions.

In mammals, C/EBPα and C/EBPβ are subject to translational control. These bZip proteins play important roles in the differentiation of a number of cell types, including hepatocytes and adipocytes, activating genes involved in gluconeogenesis and lipogenesis. The abilities of C/EBPα and C/EBPβ to stimulate transcription is modulated by the relative expression of alternative isoforms that are produced by differential translation of an mRNA. Both C/EBPα and C/EBPβ are synthesized as either a full-length isoform with potent stimulatory activity or an N-terminally truncated isoform that functions as either a weak activator or a repressor, depending on promoter context. The production of truncated isoforms from internal start codons depends on the presence of small uORFs that delay the reinitiation of translation. These uORFs are conserved across species and may allow translation to respond to cellular eIF2 activity, in a similar manner to GCN4 mRNAs; this would provide a mechanism of coupling C/EBP function to growth conditions that control the phosphorylation state of eIF2. In differentiating adipocytes, insulin activates eIF2 and causes a rapid increase in the proportion of full-length C/EBPα and C/EBPβ. The relative proportions of full-length and truncated C/EBPβ are also subject to developmental regulation. In liver cells at birth, full-length C/EBPβ is three-fold more abundant than the truncated form, but this ratio rises to 15 in the adult, thereby increasing the transcription of C/EBP target genes.

Feedback control

Many transcription factors function in an autoregulatory capacity to control the expression of their own genes in either a positive or a negative feedback loop. Positive feedback provides a mechanism for maintaining existing conditions, such as a determined or differentiated state. By contrast, negative feedback allows a transitory response that switches itself off once its purpose has been achieved. Autoregulatory circuits can be either simple or complex in character. A simple feedback loop consists of a single transcription factor that regulates the expression of its own gene, perhaps by direct binding to its promoter. A complex autoregulatory circuit might involve multiple factors that control each other's genes, thereby indirectly affecting their own synthesis.

An example of a simple positive feedback loop is provided by the POU domain factor Pit-1 that is expressed specifically in the anterior pituitary gland. Studies with transgenic mice have shown that transcription of the *pit-1* gene is dependent on Pit-1-binding sites in a tissue-specific enhancer. This autoregulation is needed to maintain expression of the *pit-1* gene once it has become activated during development under the influence of morphogen-responsive DNA elements.

A relatively simple example of negative feedback is provided by WT1, which represses transcription of its own gene. WT1 protein also inhibits the synthesis of PAX-2, a transcription factor that activates the *WT1* gene, thereby reinforcing the autoregulatory loop (Fig. 9.4). This negative feedback control may explain why WT1 is produced only transiently during the development of many tissues.

A complex positive autoregulatory circuit involves the myogenic factors that act in a combinatorial fashion to activate the expression of a range of tissue-specific promoters in the skeletal muscle lineage. Four related basic helix-loop-helix (bHLH) proteins, called MyoD, myogenin, Myf5 and MRF, perform key functions in a regulatory pathway that determines the identity of myoblasts and controls their differentiation. Each of these factors is capable of inducing the entire programme of skeletal myogenesis when introduced into

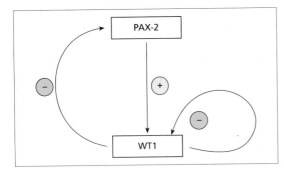

Fig. 9.4 Transcription of the *WT1* gene is subject to negative feedback control that operates at two levels. The WT1 protein inhibits expression of its own gene and also that of PAX-2, an activator of the *WT1* promoter.

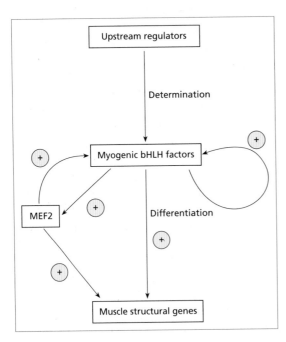

Fig. 9.5 Positive feedback during myogenesis. The initial events in skeletal muscle formation involve the determination of mesodermal precursor cells to the myogenic lineage, which expresses the bHLH proteins MyoD, myogenin, MRF4 and Myf5. These factors stimulate transcription of their own genes in a positive feedback loop. They also activate the genes encoding MEF2, which provides additional autoregulation by synergistically increasing transcription of the genes for the myogenic bHLH proteins. MEF2 also cooperates with the myogenic bHLH factors in activating a range of muscle-specific genes, such as desmin, troponin C, myosin light chain-1/3 and muscle creatine kinase.

certain types of non-muscle cell, such as the 10T1/2 fibroblast line. Recognition motifs (called E boxes) for the myogenic bHLH proteins are found in the regulatory regions of many genes that are important for skeletal muscle formation, such as the muscle creatine kinase gene. They are also found in the genes that code for the myogenic bHLH factors themselves, providing positive feedback once transcription has commenced (Fig. 9.5). Other important targets that are activated by MyoD and its relatives are genes encoding the **myocyte enhancer binding factor 2** (MEF2) family. Biochemical and genetic experiments have shown that MEF2 proteins are required as coactivators to assist the myogenic bHLH proteins in driving muscle formation. *Drosophila* mutants that lack MEF2 produce embryos that are completely bereft of skeletal, cardiac and visceral muscle. This type of combinatorial control may be very typical in the regulation of cell type-specific transcription.

MEF2 proteins act synergistically with the myogenic bHLH proteins to stimulate transcription of many tissue-specific genes in skeletal muscle cells, including the muscle creatine kinase and myosin light chain-1/3 genes. Furthermore, MyoD, myogenin and MRF4 promoters contain functional binding sites for MEF2. In cotransfection assays, MEF2 operates synergistically with myogenin or MyoD to activate the mouse MRF4 promoter. This cooperativity involves physical interaction between the bHLH domain of myogenin or MyoD and the DNA-binding domain of MEF2. Because MEF2 synthesis is induced by the myogenic bHLH proteins, its ability to activate the genes that encode these factors provides an additional wave of positive feedback that helps to maintain and amplify their expression (Fig. 9.5).

CREM provides an interesting example of negative feedback control. The promoter upstream of the first exon (P1) is insensitive to cAMP levels, and gives rise to transcripts that encode the α, β, γ and τ isoforms of CREM. However, a cAMP-inducible promoter (P2) lies between the seventh and eighth exons and drives the synthesis of a series of short, alternatively spliced mRNAs encoded by the last four exons of the gene (Fig. 9.3). The polypeptide products of these transcripts are referred to as **i**nducible **c**AMP **e**arly **r**epressors or ICERs. They contain little besides a DNA-binding domain and constitute some of the smallest known transcription factors, with a predicted mass of only 13.4 kDa in one case. Nevertheless, they bind to CREs and heterodimerize efficiently with CREM and CREB proteins. As such, they serve as potent inhibitors of cAMP-inducible transcription. The cAMP response is therefore subject to negative feedback. CREM and CREB are synthesized constitutively, but only activate transcription after being phosphorylated by protein kinase A in response to cAMP. The P2 promoter contains two CRE binding sites for CREM and CREB and so is induced by cAMP (see Chapter 5). This leads to the production of ICERs, which antagonize the action of CREM and CREB and thereby attenuate the response to cAMP. One corollary of this is that ICERs also act on the P2 promoter to inhibit their own synthesis. In the pineal gland, the P2 promoter is activated cyclically in response to rhythmic adrenergic signals that are sent by the suprachiasmatic nucleus; ICER transcripts are especially abundant at night in this endocrinal tissue.

Further reading

Reviews

Arnold, H.-H. & Winter, B. (1998) Muscle differentiation: more complexity to the network of myogenic regulators. *Curr. Opin. Genet. Dev.* **8**: 539–544.

Calkhoven, C. F. & Ab, G. (1996) Multiple steps in the regulation of transcription-factor level and activity. *Biochem. J.* **317**: 329–342.

De Cesare, D., Fimia, G. M. & Sassone-Corsi, P. (1999) Signalling routes to CREM and CREB: plasticity in transcriptional activation. *Trends Biochem. Sci.* **24**: 281–285.

Englert, C. (1998) WT1—more than a transcription factor? *Trends Biochem. Sci.* **23**: 389–393.

Foulkes, N. S. & Sassone-Corsi, P. (1992) More is better: activators and repressors from the same gene. *Cell* **68**: 411–414.

Hinnebusch, A. G. (1997) Translational regulation of yeast *GCN4*. *J. Biol. Chem.* **272**: 21661–21664.

Molkentin, J. D. & Olson, E. N. (1996) Combinatorial control of muscle development by basic helix-loop-helix and MADS-box transcription factors. *Proc. Natl Acad. Sci. USA* **93**: 9366–9373.

Wolffe, A. P. & Brown, D. D. (1988) Developmental regulation of two 5S ribosomal RNA genes. *Science* **241**: 1626–1632.

Selected papers

Regulation of genes encoding transcription factors

Duncan, S. A., Navas, M. A., Dufort, D., Rossant, J. & Stoffel, M. (1998) Regulation of a transcription factor network required for differentiation and metabolism. *Science* **281**: 692–695.

Molkentin, J. D., Black, B. L., Martin, J. F. & Olson, E. N. (1995) Cooperative activation of muscle gene expression by MEF2 and myogenic bHLH proteins. *Cell* **83**: 1125–1136.

Naidu, P. S., Ludolph, D. C., To, R. R., Hinterberger, T. J. & Konieczny, S. F. (1995) Myogenin and MEF2 function synergistically to activate the MRF4 promoter during myogenesis. *Mol. Cell. Biol.* **15**: 2707–2718.

Yee, S. P. & Rigby, P. W. J. (1993) The regulation of myogenin gene expression during the embryonic development of the mouse. *Genes Dev.* **7**: 1277–1289.

Alternative splicing of transcripts encoding transcription factors

Bickmore, W. A., Oghene, K., Little, M. H. *et al.* (1992) Modulation of DNA binding specificity by alternative splicing of the Wilms' tumour *wt1* gene transcript. *Science* **257**: 235–237.

Foulkes, N. S., Mellstrom, B., Benusiglio, E. & Sassone-Corsi, P. (1992) Developmental switch of CREM function during spermatogenesis: from antagonist to activator. *Nature* **355**: 80–84.

Foulkes, N. S., Schlotter, F., Pevet, P. & Sassone-Corsi, P. (1993) Pituitary hormone FSH directs the CREM functional switch during spermatogenesis. *Nature* **362**: 264–267.

Larsson, S. H., Charlieu, J.-P., Miyagawa, K. *et al.* (1995) Subnuclear localization of WT1 in splicing or transcription factor domains is regulated by alternative splicing. *Cell* **81**: 391–401.

Translational control of transcription factor synthesis

Calkhoven, C. F., Bouwman, P. R. J., Snippe, L. & Ab, G. (1994) Translation start site multiplicity of the CCAAT/enhancer binding protein α mRNA is dictated by a small 5′ open reading frame. *Nucleic Acids Res.* **22**: 5540–5547.

Dever, T. E., Feng, L., Wek, R. C. *et al.* (1992) Phosphorylation of initiation factor 2α by protein kinase GCN2 mediates gene-specific translational control of GCN4 in yeast. *Cell* **68**: 585–596.

Feedback control

Molina, C. A., Foulkes, N. S., Lalli, E. & Sassone-Corsi, P. (1993) Inducibility and negative autoregulation of CREM: an alternative promoter directs the expression of ICER, an early response repressor. *Cell* **75**: 875–886.

Chapter 10: Regulation of Transcription Factor Localization

The evolution of a nuclear envelope provided eukaryotes with a means for regulating gene expression that was not available to their prokaryotic progenitors. Like all proteins, transcription factors are synthesized in the cytoplasm; in order to function, they must enter the nucleus and interact with genes. For most proteins, transport into the nucleus is an active process, although the nuclear pore complex is permeable to molecules as large as 40–60 kDa. Karyophilic proteins contain one or more nuclear-localization signal (NLS) sequences that bind to receptor proteins located at the nuclear pore. Polypeptides without an NLS can often gain entry by interacting with a karyophilic partner. Once bound to the NLS receptor, proteins are translocated through the pore complex in an energy-dependent process. Nuclear uptake can be controlled in a variety of ways, and in some cases has been found to respond to environmental stimuli, cell cycle progression or developmental cues. The mechanism of regulation can involve phosphorylation at or near the NLS or protein–protein interactions that mask the NLS. In some cases, proteins are excluded from the nucleus by anchorage to a cytoplasmic protein. Such mechanisms can provide an important control point for regulating transcription.

Regulation of nucleocytoplasmic transport by phosphorylation

Chapter 8 described how the *Saccharomyces cerevisiae* bHLH factor Pho4 provokes changes in the chromatin structure of the *PHO5* promoter in response to phosphate starvation; these changes accompany a transcriptional induction of more than 100-fold. *PHO5* encodes an acid phosphatase that is secreted in order to scavenge phosphate from the yeast's extracellular environment. Pho4 also regulates a number of other genes that are induced when phosphate becomes scarce. Such control mechanisms are extremely important to allow microorganisms to utilize scarce nutrients effectively.

Activation of *PHO5* expression by Pho4 can occur in the presence of protein synthesis inhibitors, which indicates that the mechanism is post-translational. *In vivo* footprinting has shown that Pho4 only occupies its cognate DNA sites under inducing conditions. When phosphate is abundant and *PHO5* is repressed, Pho4 is localized predominantly in the cytoplasm; falling phosphate levels allow it to accumulate in the nucleus, where it can regulate its target genes. This transition is controlled by the phosphorylation of Pho4 at specific serine residues; dephosphorylation occurs under starvation conditions and correlates with nuclear entry. Phosphorylation of a serine within the NLS of Pho4 prevents its interaction with nuclear import receptors, whilst phosphorylation of two additional serine residues promotes its interaction with nuclear export receptors. Substitution of these serines for non-phosphorylatable alanine residues gives a mutant factor that is concentrated in the nucleus and activates *PHO5* constitutively, irrespective of phosphate levels. The kinase responsible for controlling Pho4 is called Pho85 and is 50% identical to the cyclin-dependent kinase Cdc28. To exert its catalytic activity, Pho85 needs to bind to a cyclin-like molecule called Pho80. Artificial overexpression of Pho80 can compromise the ability of Pho4 to activate *PHO5*. Conversely, Pho4 is found within the nucleus constitutively in strains lacking Pho80 or Pho85. Under physiological conditions, the activity of the Pho80/Pho85 kinase decreases in response to phosphate starvation, thereby allowing Pho4 to enter the nucleus and stimulate expression of its target genes (Fig. 10.1).

The mechanism controlling Pho4 localization has close parallels with the cell cycle regulation of SWI5, also in *S. cerevisiae*. As described in Chapter 8, SWI5 is a zinc finger-containing DNA-binding factor that activates the gene encoding endonuclease HO, which is required for mating type switching. The cyclin-dependent kinase Cdc28 phosphorylates SWI5 at three serine residues adjacent to its NLS. These phosphorylations inhibit the nuclear import of SWI5, as shown by site-directed mutagenesis; substitution to non-phosphorylatable residues generates a constitutively nuclear SWI5 mutant. The kinase activity of Cdc28 is cell cycle regulated; it is inactivated at the end of mitosis, whereupon SWI5 becomes dephosphorylated, enters the nucleus and activates transcription.

Cyclin-dependent kinases have also been found to regulate nuclear transport in higher organisms. Indeed, the first evidence that nucleocytoplasmic transport can be regulated reversibly by phosphorylation came from studies of the T antigen oncoprotein of simian virus 40. A site near the NLS of T antigen is phosphorylated specifically by the cyclin-dependent kinase Cdc2 and this inhibits nuclear import. By contrast, the rate of nuclear uptake is enhanced if T antigen is phosphorylated by **c**asein **k**inase **II** (CKII) at a distinct site flanking the NLS. Various other transcription factors, such as p53 and c-Myc, also contain target sites for CKII or Cdc2 in the vicinity of their NLS. In several cases, phosphorylation by cyclin-dependent kinases has been found to inhibit nuclear import whereas CKII has a stimulatory effect. It has therefore been speculated that these kinases may provide a general mechanism for the dual control of nuclear transport.

Anchorage in the cytoplasm—Rel proteins

Transcription factors of the Rel/NF-κB family play a major role in the immune response in higher organisms. Almost all the stimuli that activate lymphoid cells also induce NF-κB function; examples include cytokines such as interleukin

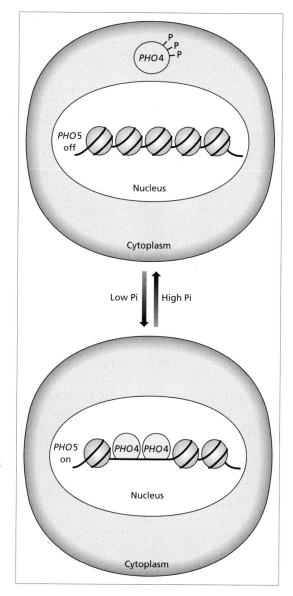

Fig. 10.1 The subcellular localization of Pho4 is controlled by phosphorylation in response to the availability of phosphate. When phosphate is abundant, Pho4 is retained in the cytoplasm due to its phosphorylation by the Pho80/Pho85 kinase complex. Phosphate starvation results in a decrease in the activity of this kinase; this allows Pho4 to accumulate in the nucleus and activate target genes such as *PHO5*.

1 (IL-1), viral infection and antigen-induced crosslinking of T and B cells. These factors are also activated by calcium ionophores, phorbol esters, free radicals and UV irradiation, and have been implicated in the pathogenesis of several tumour types, particularly of haematopoietic origin. Targets for regulation by the Rel/NF-κB proteins include genes encoding proteins that are important for the immune system (e.g. cytokines and their receptors), cell adhesion molecules (e.g. **i**ntracellular **a**dhesion **m**olecule **1**, ICAM-1) and transcription factors (e.g. c-Myc). They are also involved in the control of several viruses, such as HIV-1. NF-κB was originally believed to be a

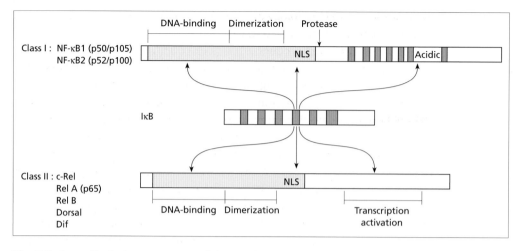

Fig. 10.2 Generalized primary structures of class I and class II Rel family members and an IκB polypeptide. The Rel homology domain is shaded pink. Ankyrin repeats are shaded red. Locations of the nuclear localization signal (NLS) and protease cleavage site are indicated, as are the regions responsible for DNA binding, dimerization and transcriptional activation. Also indicated by arrows are the regions of the Rel proteins that are thought to interact with IκB.

lymphoid-specific factor, because it is constitutively present in the nuclei of mature B cells. However, subsequent studies revealed that it is expressed in virtually all cells, but is generally sequestered in the cytoplasm; activating stimuli provoke the nuclear import of NF-κB.

Mammals contain at least five members of the Rel/NF-κB family, whereas *Drosophila* have at least two. The defining feature is the Rel homology domain, a conserved region of approximately 300 amino acids which contains the NLS and is responsible for both DNA-binding and dimerization. The vertebrate Rel polypeptides can either homodimerize or heterodimerize to generate a range of complexes with distinct DNA recognition specificities. This allows for substantial combinatorial diversity. Thus the consensus Rel recognition sequence (κB site) consists of a degenerate decamer composed of two half-sites, which reflect the specificities of the individual monomers. X-ray crystallography shows that the NF-κB1 p50 homodimer wraps itself around the DNA in a structure reminiscent of a butterfly. The Rel homology domains make multiple contacts with the κB site and consequently have very high affinity for DNA. The Rel family is divided into two classes, according to the sequences that follow the Rel homology domain (Fig. 10.2). Class 1 consists of p105 (NF-κB1) and p100 (NF-κB2); these proteins have long C-terminal domains that contain seven copies of a 33-amino-acid protein–protein interaction motif called the ankyrin repeat. The class II Rel proteins include c-Rel, RelA (p65) and RelB in mammals, Dorsal and Dif in *Drosophila*; their C-terminal regions are generally unrelated, but contain transcription activation domains that are often acidic; the Rel homology domains in this class also contain a 25-residue deletion compared with those of class I.

Most vertebrate Rel proteins are expressed at highest levels in lymphoid cells and activate expression of genes involved in immunoregulatory responses, such as those encoding interferon, cytokines and cytokine receptors. Knockout mice carrying a homozygous inactivation of the gene for p105 (NF-κB1) have a defective immune response. Rel complexes are active and located in the nucleus constitutively in mature B cells. However, they are retained in the cytoplasm in most cell types due to interaction with proteins called IκB (for inhibitor of NF-κB). There is a family of IκB proteins, the most important of which are IκBα, IκBβ and IκBε in vertebrates and Cactus in *Drosophila*. All contain five to eight copies of the ankyrin repeat, bounded by sequences of variable length. X-ray analyses of cocrystals show that the ankyrin repeats of IκBα make extensive contacts with the Rel homology domains of both subunits of a p50/p65 NF-κB heterodimer. These interactions prevent NF-κB from binding to DNA. They also mask or distort the NLS of Rel proteins, thereby sequestering NF-κB in the cytoplasm, away from its genetic targets. Individual IκBs have distinct specificities for Rel family members, thereby increasing the complexity of the regulatory network. For example, IκBγ inhbits homodimers of NF-κB1 and RelA, but alternative splicing can generate isoforms with different activities. Furthermore, deletion of three residues from the first ankyrin repeat prevents IκBγ from binding RelA but does not stop it binding to NF-κB1.

The stimuli which activate Rel factors trigger the rapid degradation of IκB. The first step involves phosphorylation of IκB polypeptides at two serine residues near the N-terminus. This can be achieved by an inducible kinase complex called IKK (for IκB kinase) that is at least 500 kDa in size and contains several components; two IKK subunits are protein kinases with leucine zippers and HLH domains. IκB can also be phosphorylated at the same sites *in vivo* by other inducible kinases, such as pp90^{rsk}. Different kinases may therefore mediate the response to distinct stimuli. Much effort is being invested in identifying the relevant signal transduction pathways, because drugs that can interfere specifically with these may prove valuable for modulating the immune response. Phosphorylation of IκB is not sufficient to dissociate it from the Rel proteins. Instead, it provides a signal for IκB to become ubiquitinated, which in turn leads to its degradation by the proteasome, a multisubunit protease complex. This releases Rel dimers to enter the nucleus, bind to target promoters and enhancers and activate gene expression.

NF-κB1 and NF-κB2 are synthesized as large polypeptides (p105 and p100, respectively) that are anchored in the cytoplasm even without the help of IκB. Inducing stimuli trigger the proteolytic processing of p105 and p100 so as to remove the C-terminal halves of these proteins, which are responsible for cytoplasmic retention, thereby releasing active p50 and p52, respectively. In fact, IκBγ comprises the C-terminal ankyrin repeat domain of p105, expressed from a subgenomic mRNA. As with IκB, proteolysis of p105 and p100 involves phosphorylation followed by ubiquitination. This process is slower than the degradation of IκB and may mediate a more sustained response to extracellular stimuli. Not only does it involve NF-κB1 and NF-κB2, but it also affects the class II Rel proteins, which can be precluded from nuclear entry by dimerizing with p105 and p100. Indeed, it is estimated that as much as 80% of RelA can be retained in the cytoplasm of resting cells through interaction with p105 and p100, whereas only 10–20% is held in complexes with IκB. Thus, two separate pathways contribute to the transcriptional induction of genes with κB sites (Fig. 10.3). The relative importance of these pathways will almost certainly vary according to the precise sequence of the κB site and also the nature of the inducing signal. HIV-1 encodes a protease that cleaves p105 during acute infection; this will help ensure that the viral genome is transcribed at high levels due to activation by NF-κB.

IκBα is encoded by one of the genes that is controlled by κB sites. This provides the system with a mechanism for negative feedback, whereby newly synthesized IκBα can block further gene activation by free NF-κB complexes. In addition, IκBs have been shown to displace DNA-bound Rel proteins from their κB target sites, allowing the initial wave of gene activation to be subsequently silenced once the stimulus is removed.

Members of the Rel and IκB families play important roles in embryonic morphogenesis. This was first demonstrated in *Drosophila*, where Dorsal is essential in generating the dorsoventral axis. *Dorsal⁻* embryos show a dorsalized phenotype; that is to say, they fail to form ventral structures. In the wild type, Dorsal accumulates almost exclusively in the nucleus within the ventral part of the developing blastoderm, whereas it is predominantly cytoplasmic along the dorsal side. A continuous gradient forms in between these two extremes, with the effective concentration of Dorsal being determined by its localization rather than its abundance. In dorsal regions, nuclear entry by Dorsal is prevented by an IκB family member called Cactus. More ventrally, Cactus becomes degraded in response to signals from a cell surface receptor called Toll, which has a cytoplasmic domain that is related to the mammalian IL-1 receptor. Toll, in turn, responds to extracellular positioning cues that are derived from the mother fly. In *cactus⁻* mutants, Dorsal becomes concentrated in the nucleus irrespective of position, and the embryo assumes a ventralized phenotype. At Dorsal target promoters, the number and affinity of Dorsal-binding sites effectively sets a particular threshold Dorsal concentration for transcriptional activation. This limits the expression of key zygotic regulatory genes and thereby defines specific domains within the embryo. For example, one of the target genes that is regulated by this system is *twist*, which has low-affinity upstream binding sites for Dorsal and is consequently only expressed in the most ventral cells, where the nuclear concentration of Dorsal is highest. By contrast, the *rhomboid* gene has high-affinity Dorsal-binding sites and is therefore expressed laterally, despite the lower nuclear concentration of Dorsal. In

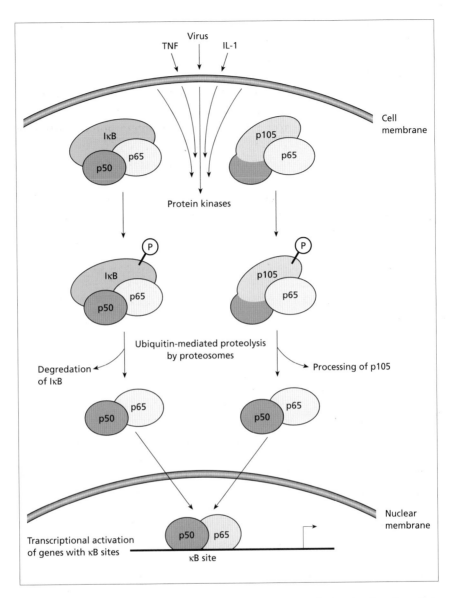

Fig. 10.3 Pathways for gene induction by Rel/NF-κB proteins. Inducing signals such as viral infection or cytokines (e.g. interleukin-1 (IL-1) and tumour necrosis factor (TNF)) activate protein kinases which phosphorylate IκB and the p105 precursor of NF-κB1. Once phosphorylated at specific sites, IκB and p105 are subject to degradation or processing, respectively, by ubiquitin-dependent proteasomes. This releases active Rel/NF-κB proteins from cytoplasmic retention signals and allows them to translocate to the nucleus and activate genes with κB DNA sites.

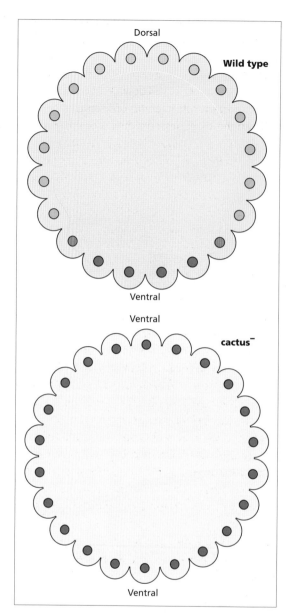

Fig. 10.4 The nucleocytoplasmic localization of Dorsal dictates dorsoventral positioning in *Drosophila* embryos. Dorsal protein is located around the periphery of the developing syncitial blastoderm. In ventral regions, it is found primarily in the nuclei, whereas it is mainly cytoplasmic in dorsalmost regions; in lateral regions it is partitioned between the nucleus and the cytoplasm. Consequently, the highest nuclear concentrations of Dorsal protein are achieved in the ventral region of the embryo. Nuclear exclusion of Dorsal is achieved by Cactus, a member of the IκB family. In mutants that lack Cactus function, Dorsal is nuclear in all regions and the embryo displays a ventralized phenotype. The diagram represents a transverse section of a *Drosophila* blastoderm; circles represent nuclei, which are located at the periphery; the regions that contain Dorsal protein are shaded.

effect Dorsal acts as a morphogen, with its nuclear concentration dictating the pattern of gene expression at each point along the dorsoventral axis (Fig. 10.4).

In developing chicks, c-Rel is expressed in undifferentiated mesoderm cells at the tips of the limb buds, where patterning signals operate. Blocking Rel/ NF-κB activity resulted in limb abnormalities, most often truncations. This was achieved by using viral vectors to express in limb buds mutant forms of IκB that cannot be phosphorylated and hence are not degraded. Expression of the chick *Twist* gene is diminished when Rel/NF-κB activation is blocked. The importance of *Twist* for vertebrate limb development has been shown using knockout mice, where the *Twist*-null embryos die midway through gestation with truncated limbs. In humans, a condition called Saethre–Chotzen syndrome, where limb abnormalities occur, has been traced to mutations in *Twist*. *Twist* itself encodes a transcription factor with an HLH domain. It is striking that the control of Twist production by Rel proteins has been conserved between flies and vertebrates, and in both cases plays an important role in embryonic morphogenesis.

Rearrangements and amplifications of genes encoding Rel proteins are associated with the pathogenesis of some cancers of the haematopoietic system, including up to 10% of human B- and T-cell leukaemias. For example, chromosomal rearrangements in some lymphoid neoplasms generate truncated versions of NF-κB2 in which the ankyrin repeats have been deleted. These C-terminally truncated mutants are constitutively nuclear in their unprocessed form, can bind DNA and are potent transactivators. Overexpression of these lymphoma-derived NF-κB2 mutants is sufficient to transform murine fibroblasts. A molecular hallmark of Hodgkin's lymphoma is constitutive nuclear NF-κB activity. One of the most potent oncoproteins known is v-Rel, the transforming product of an avian erythroblastosis virus called REV-T. This retrovirus can induce fatal lymphoma in a chicken within 1–2 weeks of infection, an extremely short latency period. v-Rel is missing 118 C-terminal residues that are present in its cellular progenitor c-Rel and contains sequences of the viral Env protein fused at each end. As a consequence of the C-terminal truncation, v-Rel is constitutively nuclear; this seems to account for its transforming potency, because deletions of the C-terminus of chicken c-Rel can increase dramatically the ability of this proto-oncoprotein to transform avian cells.

Rel proteins have also been implicated in malignant transformation by other oncogenic viruses. An example is the **human T-cell leukaemia virus** type 1 (HTLV-1), which transforms human CD4$^+$ T lymphocytes. HTLV-1 encodes a protein called Tax, which can transform T cells in culture and induce leukaemias in transgenic mice. Tax is a transcription factor which induces the expression of c-Rel. Tax-transformed leukaemic T cells display constitutive nuclear NF-κB activity, because Tax stimulates the processing of p105 and p100 and the degradation of IκBα and IκBβ. Tax has been shown to bind directly to the proteasome and also to p105; it may therefore accelerate the proteolytic maturation of p105 by recruiting it to the proteasome (Fig. 10.5). It is thought that deregulation of NF-κB causes increased production of growth factors (e.g. IL-2) and their receptors; the resultant autocrine regulation might trigger T-cell proliferation. Epstein–Barr virus and hepatitis B virus have also been shown to promote the degradation of IκB and the processing of p105.

It is widely believed that Rel/NF-κB factors may prove to be extremely useful therapeutic targets, given their importance in the immune response and in viral pathogenesis. For example, pharmaceutical inhibitors of IκB release or degradation might block the NF-κB-mediated activation of HIV-1 transcription, or the deregulation of NF-κB in certain lymphoid cancers. Antisense oligonucleotides against the RelA mRNA are reported to have potent tumour suppressive effects in a mouse model system.

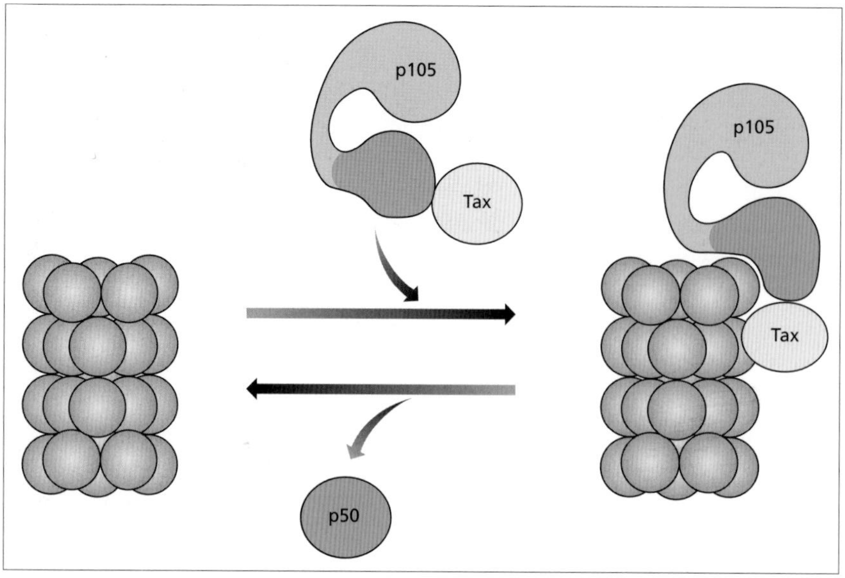

Fig. 10.5 The Tax oncoprotein of HTLV-1 is thought to recruit p105 (NF-κB1) to the 20S proteasome and thereby stimulate its processing. The 20S proteasome is a multicatalytic complex composed of 28 subunits, which are arranged into four heptameric rings that form a cylinder. Tax can bind to two of these subunits. It also binds to p105. It may therefore anchor p105 to the proteasome and thereby accelerate its processing into p50, which can enter the nucleus and activate transcription.

Nuclear entry in response to calcium—NF-AT

In addition to the Rel/NF-κB family, a full immune response requires a transcription factor called NF-AT. For example, the promoter of the important cytokine IL-2 is bound and activated by both NF-AT and NF-κB. NF-AT has been studied in great detail, because its activation is inhibited by the main immunosuppressant drugs in current clinical use, FK506 and cyclosporin A. It plays a key role in hypertrophy of cardiac and skeletal muscle, an important adaptive response to physical activity and to various pathological stimuli. Mammals have four known genes encoding NF-AT proteins, which range in size between about 700 and 1100 amino acid residues. The DNA-binding domain bears distant sequence similarity to the Rel homology domain, a relationship that is born out by crystallographic comparisons. Most of the DNA sites that are utilized by NF-AT are flanked immediately downstream by non-consensus recognition sequences for AP-1 (a generic term for the various combinations of Fos and Jun dimers). Neither NF-AT nor AP-1 binds tightly to these composite sites, but together they display strong cooperativity due to extensive protein–protein interactions between their DNA-binding domains. The presence of NF-AT causes a 10-fold increase in the affinity of AP-1 for such sites.

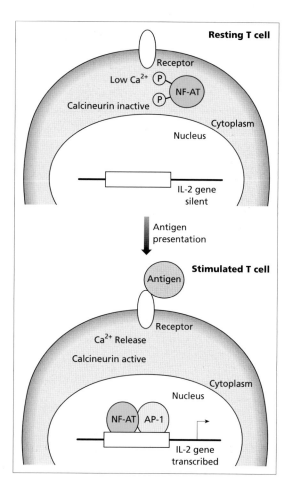

Fig. 10.6 Antigenic stimulation of
T-lymphocytes triggers the synthesis of AP-1
and the nuclear import of NF-AT, leading to
induction of IL-2 gene transcription. In resting
T cells, NF-AT is confined to the cytoplasm by
phosphorylation. Antigenic stimulation triggers
the release of calcium and the synthesis of
AP-1. Calcium activates the calmodulin-
dependent phosphatase calcineurin, which
dephosphorylates NF-AT and thereby allows it
to enter the nucleus. NF-AT and AP-1 bind
cooperatively to the IL-2 promoter and induce
transcription.

NF-AT is largely cytoplasmic in resting T cells. Unlike the previous examples, where nuclear uptake was triggered by protein kinases, phosphate
removal provides the trigger for transport of NF-AT. Stimulation of the T-cell
antigen receptors causes release of Ca^{2+} ions from intracellular stores. This
activates calmodulin, which provokes a Ca^{2+}-dependent protein phosphatase
called calcineurin to dephosphorylate residues in the N-terminus of NF-AT,
thereby uncovering its NLS and allowing it to enter the nucleus. The immuno-
suppressants cyclosporin A and FK506 block this step by inhibiting calcineurin. T-cell activation is also accompanied by an increase in the synthesis
of AP-1 proteins (Fos and Jun), which cooperate with NF-AT to induce the
transcription of genes such as *IL-2* (Fig. 10.6). Such a combinatorial system
will increase the specificity of the response. It also serves to integrate regulatory
inputs from separate signalling pathways, because NF-AT and AP-1 respond to
different sets of stimuli.

The four genes known to encode NF-AT polypeptides in mammals
exhibit different patterns of expression and appear to perform distinct roles.

Disruption of the murine gene encoding NF-ATp results in a defective immune response. By contrast, genetic disruption of another family member, called NF-ATc, causes mice to die before day 15 of gestation due to a failure to form the aortic and pulmonary valves. This abnormality correlates with a burst of NF-ATc activation that occurs in the developing hearts of normal mice. NF-ATp appears not to be expressed in cardiac tissue and the heart develops normally in *NF-ATp* knockouts. One of the defects caused by disruption of NF-ATc in mice is strikingly similar to a class of human congenital heart diseases. It has therefore been suggested that identifying the upstream activators and target loci of NF-ATc in the developing heart may help reduce the frequency of these human birth disorders. This might have a considerable impact, because fetal heart malformations are implicated in many stillbirths and spontaneous abortions, whilst congenital heart defects are found in 1% of live births. Another family member that is expressed in the heart is NF-AT3. Instead of AP-1, NF-AT3 interacts synergistically with a zinc finger protein called GATA4 to stimulate the expression of various genes involved in the hypertrophic response. Transgenic mice that express constitutively activated forms of calcineurin or NF-AT3 develop cardiac hypertrophy and heart failure that mimic human heart disease.

STATs—transcription factors that bind directly to cell-surface receptors

The STATs (**s**ignal **t**ransducers and **a**ctivators of **t**ranscription) are a family of transcription factors that exist in a latent form in the cytoplasm. They become activated following phosphorylation on tyrosine in response to a range of extracellular stimuli. An unusual feature of these transcription factors is that they are activated directly at the plasma membrane, rather than by a cascade of cytoplasmic kinases. Once phosphorylated, they leave the plasma membrane and enter the nucleus in order to regulate a number of important target genes.

The mammalian STATs are 750–850 amino acids long, with a central DNA-binding domain, an **S**rc **h**omology 2 (SH2) domain and a C-terminal activation domain (Fig. 10.7). They are encoded by seven genes, which are located on three chromosomal clusters and appear to have evolved by consecutive

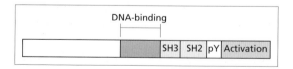

Fig. 10.7 Generalized primary organization of STAT family members. The DNA-binding domain is found between residues ~400 and ~500. It is followed by an Src homology 2 (SH2) domain. A less conserved region with weak homology to an SH3 domain lies in between the DNA-binding and SH2 domains. All activated STATs contain phosphotyrosine (pY) in the vicinity of residue 700. A transcription activation domain is found at the C-terminus.

duplications of a primordial precursor. Differential splicing adds to the variety, giving at least a dozen polypeptides from the seven STAT genes. Although they were originally identified as mediating the interferon response, it is now clear that STATs become activated in cells treated with a wide range of extra-cellular signalling polypeptides. In many cases this is mediated by the **Ja**nus **k**inases (JAKs), a group of large (~1200-residue) tyrosine kinases that bind to cytokine receptors for interferons, interleukins, growth hormone, prolactin and other polypeptides. STATs can also be activated by receptors with in-trinsic, ligand-dependent tyrosine kinase activity, for example the **e**pidermal **g**rowth **f**actor (EGF) and **p**latelet-**d**erived **g**rowth **f**actor (PDGF) receptors. Ligand binding causes the receptors to dimerize, leading to reciprocal tyrosine phosphorylation either by the intrinsic kinase domains or the associated JAKs. Phosphotyrosines on the cytoplasmic tails of the receptors then serve as dock-ing sites for the SH2 domains of STATs. Once docked, the STAT itself under-goes phosphorylation on a single tyrosine around residue 700; this activates it, causing it to form homo- or heterodimers. Dimerization is dependent on the intermolecular interaction of the SH2 domains with the partner's phosphoty-rosine. Phosphopeptides that include Tyr-701 of STAT1 will competitively block homodimerization or association with STAT2. Unlike the monomers, STAT dimers can bind to DNA after translocation to the nucleus (Fig. 10.8).

STAT activation displays very high cytokine specificity. For example, IL-2 activates STAT5, IL-4 activates STAT6 and IL-12 activates STAT4 in lympho-cytes. This specificity is achieved because the interaction between receptor phosphotyrosines and STAT SH2 domains can discriminate between indi-vidual family members. Swapping the SH2 domains between STATs 1 and 2 reverses their cytokine response profiles. By contrast, JAKs appear not to dis-criminate between different STATs. As a result of the very high specificity that is achieved in the receptor/STAT interactions, individual STAT family members perform very precise physiological roles. For example, knockout of the *STAT1* gene by homologous recombination results in viable mice that cannot respond to viral or bacterial infections; this presumably reflects the importance of STAT1 for the interferon response. STAT4 is activated in T cells exposed to IL-12, a cytokine that stimulates T helper 1 ($T_H 1$) cell development; STAT4 knockout mice are deficient in $T_H 1$ cells, but otherwise normal. STAT6 is only activated by IL-4, which promotes the development of $T_H 2$ cells; accordingly, the STAT6 knockout mice lack $T_H 2$ cell function. The phenotypes of all seven STAT knockouts indicate that each family member has a unique and crucial role in mice.

STAT activation is generally transient; the maximal response to interferon-γ stimulation takes 20–30 minutes, and subsequent downregulation occurs in a similar interval once tyrosine kinase activity ceases. It can be prolonged by treatment of cells with inhibitors of phosphotyrosine phosphatases, suggesting that dephosphorylation is important for downregulating the response. Pulse-labelling experiments have shown that protein degradation is not responsible

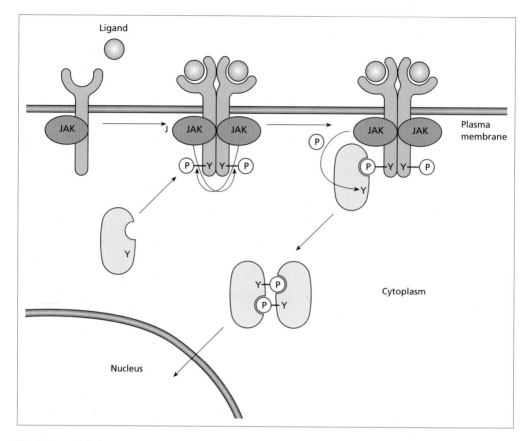

Fig. 10.8 Model of STAT activation. Ligand binding causes cell surface receptors to aggregate. This brings the associated Janus kinases (JAKs) into sufficient proximity to allow transphosphorylation of the receptor chains at multiple sites. Phosphotyrosine residues on the internal domain of the receptor serve to recruit STATs via their SH2 domains. Once bound, the STATs become phosphorylated by the associated JAKs. Phosphorylated STATs dissociate from the receptor and dimerize due to the intermolecular association of their SH2 domains with a phosphotyrosine in the vicinity of residue 700. Having dimerized, the STATs enter the nucleus and activate transcription.

for clearing STATs from their DNA targets. It is probable that dephosphorylation of tyrosine causes the dimers to dissociate, thereby eliminating their DNA-binding potential and triggering their return to the cytoplasm.

A cofactor found in several cellular compartments— β-catenin

Like the STATs, the cofactor β-catenin can be found at the plasma membrane; here it associates with E-cadherin and α-catenin, proteins which mediate cellular adhesion. However, β-catenin also serves as an effector of the Wnt/ Wingless signal transduction pathway. When cells are exposed to a Wnt or

Wingless ligand, β-catenin enters the nucleus and binds to TCF (**T-c**ell **f**actor) and LEF (**l**ymphoid-**e**nhancer **f**actor), HMG domain-containing factors that bend DNA. This interaction is mediated by an Armadillo repeat domain in β-catenin, which forms a rigid extended superhelix with a long positively charged groove that contacts acidic patches on its target proteins. The C-terminal region of β-catenin contains a transcription activation domain which allows it to serve as a cofactor to stimulate expression of genes with appropriately positioned recognition sites for TCF or LEF. The β-catenin pathway plays an important role in the embryonic development of mammals, frogs and fruitflys. For example, injecting LEF1 mRNA into the ventral side of a *Xenopus* embryo results in the formation of a secondary axis, whereas a dominant-negative version of LEF1 inhibits generation of the endogenous axis. In *Drosophila*, β-catenin is referred to as Armadillo and is required for Wingless to determine correct segment polarity; this will be described further in Chapter 14.

In the absence of Wnt or Wingless signalling, any free cytoplasmic β-catenin is unstable and rapidly degraded. This is mediated by a multicomponent complex which contains the serine/threonine protein kinase GSK3 (**g**lycogen **s**ynthase **k**inase **3**). GSK3 phosphorylates β-catenin and this earmarks it for ubiquitination and degradation by proteasomes. Indeed, β-catenin can be stabilized by substitution of its GSK3 phosphorylation sites. However, β-catenin is a poor substrate for GSK3 in the absence of a second component of the complex called axin, which binds both β-catenin and GSK3 and acts as a scaffold to facilitate phosphorylation. Axin fragments that lack individual protein-binding domains cause β-catenin levels to rise and have potent dominant-negative effects when introduced into mammalian cells or the ventral side of *Xenopus* embryos. When cells are stimulated by Wnt or Wingless, the axin-containing complex is inhibited, allowing free unphosphorylated β-catenin to accumulate, enter the nucleus and stimulate transcription (Fig. 10.9).

Another component of the axin/GSK3 complex is the tumour suppressor APC (**a**denomatous **p**olyposis **c**oli). Although the function of APC is unclear, its loss results in increased pools of free cytoplasmic β-catenin. APC is found to be mutated in many cases of colon cancer, allowing constitutively active TCF/β-catenin complexes to form in nuclei. APC-positive colon carcinomas sometimes carry dominant mutations in the GSK3 phosphorylation sites of β-catenin, enabling it to accumulate in nuclear complexes with TCF. Similar mutations have been found in malignant melanomas, which stabilize β-catenin and result in constitutive complexes with LEF1. The end result is the same for loss-of-function mutations in APC and activating mutations in β-catenin: deregulated expression of target genes for TCF and LEF factors. One of the key targets is the proto-oncogene *cyclin D1*, which has TCF sites in its promoter. Abnormal levels of β-catenin cause the overproduction of cyclin D1, which in turn drives cell proliferation. This may constitute an important step towards neoplastic transformation in many colon tumours and other types of malignancy.

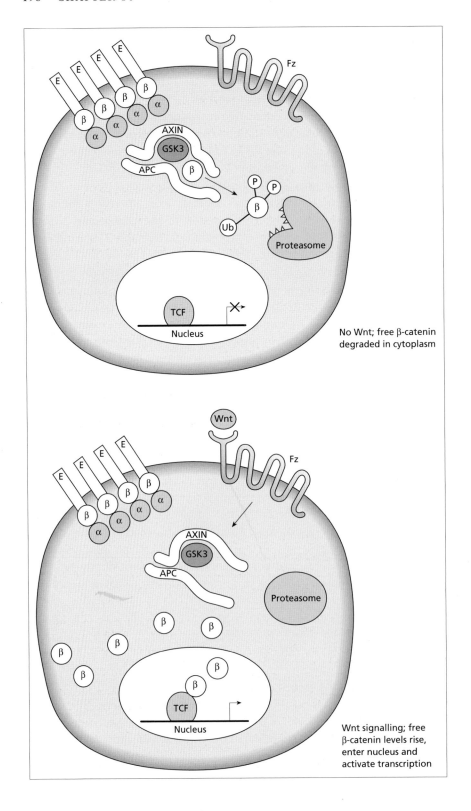

No Wnt; free β-catenin
degraded in cytoplasm

Wnt signalling; free
β-catenin levels rise,
enter nucleus and
activate transcription

Further reading

Reviews

Baeuerle, P. A. (1998) IκB-NF-κB structures: at the interface of inflammation control. *Cell* **95**: 729–731.

Baeuerle, P. A. & Baltimore, D. (1996) NF-κB: 10 years after. *Cell* **87**: 13–20.

Bienz, M. (1999) APC: the plot thickens. *Curr. Opin. Genet. Dev.* **9**: 595–603.

Clevers, H. & van de Wetering, M. (1997) TCF/LEF factors earn their wings. *Trends Genet.* **13**: 485–489.

Crabtree, G. R. (1999) Generic signals and specific outcomes: signalling through Ca^{2+}, calcineurin, and NF-AT. *Cell* **96**: 611–614.

Darnell, J. E., Jr (1997) STATs and gene regulation. *Science* **277**: 1630–1635.

Darnell, J. E., Kerr, I. M. & Stark, G. R. (1994) Jak-STAT pathways and transcriptional activation in response to IFNs and other extracellular signalling proteins. *Science* **264**: 1415–1421.

Ghosh, S., May, M. J. & Kopp, E. B. (1998) NF-κB and Rel proteins: evolutionarily conserved mediators of immune responses. *Annu. Rev. Immunol.* **16**: 225–260.

Ihle, J. N. (1996) STATs: signal transducers and activators of transcription. *Cell* **84**: 331–334.

Lenburg, M. E. & O'Shea, E. K. (1996) Signalling phosphate starvation. *Trends Biochem. Sci.* **21**: 383–387.

Neumann, M., Marienfeld, R. & Serfling, E. (1997) Rel/NF-κB transcription factors and cancer: oncogenesis by dysregulated transcription. *Int. J. Oncol.* **11**: 1335–1347.

Rao, A., Luo, C. & Hogan, P. G. (1997) Transcription factors of the NFAT family: regulation and function. *Annu. Rev. Immunol.* **15**: 707–747.

Stancovski, I. & Baltimore, D. (1997) NF-κB activation: the IκB kinase revealed? *Cell* **91**: 299–302.

Thanos, D. & Maniatis, T. (1995) NF-κB: a lesson in family values. *Cell* **80**: 529–532.

Vandromme, M., Gauthier-Rouviere, C. & Fernandez, A. (1996) Regulation of transcription factor localization: fine-tuning of gene expression. *Trends Biochem. Sci.* **21**: 59–64.

Selected papers

Pho4

Komeili, A. & O'Shea, E. K. (1999) Roles of phosphorylation sites in regulating activity of the transcription factor Pho4. *Science* **284**: 977–980.

O'Neill, E. M., Kaffman, A., Jolly, E. R. & O'Shea, E. K. (1996) Regulation of PHO4 nuclear localization by the PHO80–PHO85 cyclin–cdk complex. *Science* **271**: 209–212.

Fig. 10.9 (*opposite*) Wnt signalling allows free β-catenin to accumulate and activate transcription. In the absence of Wnt, free cytoplasmic β-catenin (β) is phosphorylated by GSK3 in a complex that contains APC and axin; the phosphorylated β-catenin is then ubiquitinated and degraded by the proteasome. When Wnt binds to its receptor, a seven-transmembrane polypeptide called Frizzled (Fz), a signal is sent which inhibits the axin-mediated phosphorylation of β-catenin by GSK3; this does not prevent the intrinsic kinase activity of GSK3, but may disrupt or dissociate the quaternary complex. This allows free β-catenin to accumulate; some of it enters the nucleus, binds to TCF or LEF and activates transcription. In addition to this role in signal transduction, β-catenin is also found in complexes at the plasma membrane with α-catenin (α) and E-cadherin (E), which mediate cellular adhesion.

SWI5

Moll, T., Tebb, G., Surana, U., Robitsch, H. & Nasmyth, K. (1991) The role of phosphorylation and the CDC28 protein kinase in cell cycle-regulated nuclear import of the *S. cerevisiae* transcription factor SWI5. *Cell* **66**: 743–758.

NF-κB

Chen, Z., Hagler, J., Palombella, V. J. *et al.* (1995) Signal-induced site-specific phosphorylation targets IκBα to the ubiquitin–proteasome pathway. *Genes Dev.* **9**: 1586–1597.

Henkel, T., Machleidt, T., Alkalay, I., Kronke, M., Ben-Neriah, Y. & Baeuerle, P. A. (1993) Rapid proteolysis of IκB-α is necessary for activation of transcription factor NF-κB. *Nature* **365**: 182–185.

Henkel, T., Zabel, U., van Zee, K., Muller, J. M., Fanning, E. & Baeuerle, P. A. (1992) Intramolecular masking of the nuclear location signal and dimerization domain in the precursor for the p50 NF-κB subunit. *Cell* **68**: 1121–1133.

Jacobs, M. D. & Harrison, S. C. (1998) Structure of an IκBα/NF-κB complex. *Cell* **95**: 749–758.

Rousset, R., Desbois, C., Bantignies, F. & Jalinot, P. (1996) Effects on NF-κB1/p105 processing of the interaction between the HTLV-1 transactivator Tax and the proteasome. *Nature* **381**: 328–331.

Dorsal

Roth, S., Stein, D. & Nusslein-Volhard, C. (1989) A gradient of nuclear localization of the *dorsal* protein determines dorsoventral pattern in the *Drosophila* embryo. *Cell* **59**: 1189–1202.

Rushlow, C. A., Han, K., Manley, J. L. & Levine, M. (1989) The graded distribution of the *dorsal* morphogen is initiated by selective nuclear transport in *Drosophila*. *Cell* **59**: 1165–1177.

Steward, R. (1989) Relocalization of the *dorsal* protein from the cytoplasm to the nucleus correlates with its function. *Cell* **59**: 1179–1188.

NF-AT

Beals, C. R., Clipstone, N. A., Ho, S. N. & Crabtree, G. R. (1997) Nuclear localization of NF-ATc by a calcineurin-dependent, cyclosporin-sensitive intramolecular interaction. *Genes Dev.* **11**: 824–834.

Luo, C., Shaw, K. T.-Y., Raghavan, A. *et al.* (1996) Interaction of calcineurin with a domain of the transcription factor NFAT1 that controls nuclear import. *Proc. Natl Acad. Sci. USA* **93**: 8907–8912.

Molkentin, J. D., Lu, J.-R., Antos, C. L. *et al.* (1998) A calcineurin-dependent transcriptional pathway for cardiac hypertrophy. *Cell* **93**: 215–228.

Shibasaki, F., Price, E. R., Milan, D. & McKeon, F. (1996) Role of kinases and the phosphatase calcineurin in the nuclear shuttling of transcription factor NF-AT4. *Nature* **382**: 370–373.

Timmerman, L. A., Clipstone, N. A., Ho, S. N., Northrop, J. P. & Crabtree, G. R. (1996) Rapid shuttling of NF-AT in discrimination of Ca^{2+} signals and immunosuppression. *Nature* **383**: 837–840.

STATs

Heim, M. H., Kerr, I. M., Stark, G. R. & Darnell, J. E., Jr (1995) Contribution of Stat SH2 groups to specific interferon signalling by the Jak-Stat pathway. *Science* **267**: 1347–1349.

Shuai, K., Horvath, C. M., Huang, L. H. *et al.* (1994) Interferon activation of the transcription factor Stat91 involves dimerization through SH2–phosphotyrosyl peptide interactions. *Cell* **76**: 821–828.

Stahl, N., Farruggella, T. J., Boulton, T. G. *et al.* (1995) Choice of STATs and other substrates specified by modular tyrosine-based motifs in cytokine receptors. *Science* **267**: 1349–1353.

Wen, Z., Zhong, Z. & Darnell, J. E., Jr (1995) Maximal activation of transcription by Stat1 and Stat3 requires both tyrosine and serine phosphorylation. *Cell* **82**: 241–250.

β-Catenin

Behrens, J., von Kries, J. P., Kuhl, M. *et al.* (1996) Functional interaction of β-catenin with the transcription factor LEF-1. *Nature* **382**: 638–642.

Korinek, V., Barker, N., Morin, P. J. *et al.* (1997) Constitutive transcriptional activation by a β-catenin-Tcf complex in APC$^{-/-}$ colon carcinoma. *Science* **275**: 1784–1787.

Morin, P. J., Sparks, A. B., Korinek, V. *et al.* (1997) Activation of β-catenin-Tcf signalling in colon cancer by mutations in β-catenin or APC. *Science* **275**: 1787–1790.

Riese, J., Yu, X., Munnerlyn, A. *et al.* (1997) LEF-1, a nuclear factor coordinating signalling inputs from wingless and decapentaplegic. *Cell* **88**: 777–787.

Tetsu, O. & McCormick, F. (1999) β-Catenin regulates expression of cyclin D1 in colon carcinoma cells. *Nature* **398**: 422–426.

van de Wetering, M., Cavallo, R., Dooijes, D. *et al.* (1997) Armadillo coactivates transcription driven by the product of the *Drosophila* segment polarity gene dTCF.*Cell* **88**: 789–799.

Chapter 11: Regulation of Transcription Factor Activity

In addition to controlling the production and intracellular localization of transcription factors, cells can also regulate gene expression by modulating the specific activity of extant factors. This provides a means of responding rapidly to extracellular stimuli. A variety of mechanisms can be utilized to achieve this, such as the binding of ligands or covalent modifications. Several different covalent modifications can occur, including ubiquitination, glycosylation, acetylation and, most commonly, phosphorylation. Transcription factors can respond to such signals in a number of ways, ranging from activation to degradation. The aim of this chapter is to describe some important examples of how transcription factor activity can be regulated.

Regulation by phosphorylation

Phosphorylation is almost certainly the most commonly used mechanism for regulating extant transcription factors. Several features of phosphorylation make it ideally suited for controlling the activity of proteins in the short term. One virtue of this mechanism is that it can be extremely rapid: for example, c-Jun becomes phosphorylated within 15 minutes after cells have been stimulated with phorbol esters. Another merit is that the modification is readily reversible, allowing kinases and phosphatases to control transcription in a highly dynamic manner. In most cases, extracellular stimuli lead to increased phosphorylation; in these cases, phosphate removal will be important in deactivating factors once the stimulus is withdrawn. For example, transcriptional attenuation following cyclic AMP (cAMP) removal requires that protein phosphatase 1 dephosphorylates cAMP-response element-binding protein (CREB) at Ser-133. Dephosphorylation can also, occasionally, be the primary response to environmental stimuli, as is seen when NF-AT becomes activated by the phosphatase calcineurin. The addition and removal of phosphates is very cheap energetically, as compared with the metabolic cost of synthesizing a new protein. Phosphorylation is also very versatile. For example, a particular kinase can sometimes elicit distinct responses from different transcription factor substrates. Conversely, several functions of the same factor might be controlled by different kinases, depending upon the positions of phosphorylated residues. Because of this last feature, phosphorylation can be perfectly suited for integrating information from separate signal transduction pathways.

A very good example of how several kinase cascades can converge on a single promoter site is provided by the **s**erum **r**esponse **e**lements (SREs). These DNA motifs are found upstream or downstream of many genes that are

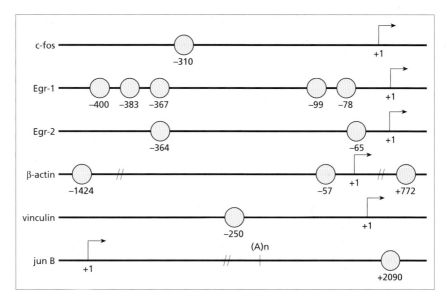

Fig. 11.1 Locations of serum response elements (SREs) in several growth factor-responsive 'immediate early' genes. SREs are shown by circles and their positions relative to the transcription start site (+1) are indicated. (A)n denotes the polyadenylation site of the *jun-B* gene.

induced rapidly when quiescent cells are stimulated with serum (Fig. 11.1). Examples of such 'immediate early' genes include *c-fos*, *jun-B* and *β-actin*. The prototype SRE is found in the *c-fos* promoter, centred 310 bp upstream of the start site. *In vivo* footprinting has shown that this SRE is occupied constitutively by two proteins, SRF (**s**erum **r**esponse **f**actor) and a member of a family called TCF (**t**ernary **c**omplex **f**actor; not to be confused with the HMG domain-containing **T**-**c**ell **f**actor (TCF) family that was encountered in previous chapters!) (Fig. 11.2). An SRF dimer binds with high affinity to the *c-fos* SRE and recruits TCFs, which bind inefficiently on their own. This functional interaction with SRF was exploited in order to isolate members of the TCF family, using a technique that screens for protein–protein interactions in living yeast cells (Box 11.1). An activation domain within SRF is unmasked by its association with TCFs. However, the response to growth factors in serum is mediated primarily by the TCFs. These stimulate transcription following phosphorylation of a cluster of sites in their activation domains by at least three distinct groups of **m**itogen-**a**ctivated **p**rotein **k**inases (MAPKs).

The best characterized MAPK pathway involves ERK1 and ERK2; it plays an important part in the response to many growth factors. Both receptor tyrosine kinases and G protein-coupled serpentine receptors feed into this signalling cascade (Fig. 11.3). It also involves the oncoprotein Ras, which is deregulated by mutations in many types of cancer. The pathway leads to the MAPKs ERK1 and ERK2; once these have become phosphorylated, they translocate from the cytoplasm to the nucleus and activate TCFs. The other

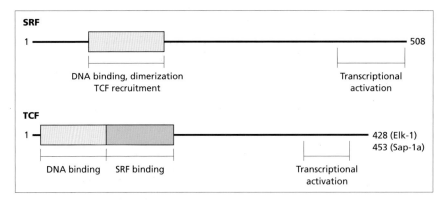

Fig. 11.2 Primary organization of SRF (serum response factor) and the TCFs (ternary complex factors) Elk-1 and SAP-1a. The DNA-binding domain of SRF contains a conserved motif called the MADS box which is also found in certain yeast and plant transcription factors. The TCFs Elk-1 and Sap-1a share a very similar organization. Both have an N-terminal DNA-binding domain of a type that is found in at least six groups of transcription factors and is referred to as the ETS domain. Adjacent to this is a region required for interaction with SRF. A region near the C-terminus is responsible for transcriptional activation.

Box 11.1 Yeast genetic screens for isolating novel proteins that bind to a known protein

The technique is illustrated by describing an example of its use, the isolation of factors that interact with human SRF. A yeast strain was engineered that carried a *lacZ* reporter gene driven by a promoter with an SRE sequence that is bound by SRF. The strain also carried a plasmid that allowed galactose-inducible expression of human SRF. Although SRF can bind the SRE in yeast, it is unable to stimulate transcription due to an absence of accessory factors. In order to isolate human proteins that bind to SRF, this strain was also transformed with a randomly primed HeLa cDNA library that had been subcloned downstream of the VP16 activation domain in a galactose-inducible expression vector. Following induction, VP16/cDNA fusion proteins that interact with SRF or bind directly to the SRE stimulate the *lacZ* reporter to express β-galactosidase, which turns the yeast blue. cDNAs that gave blue colonies were then recovered and characterized. The approach was used to isolate the TCF proteins Sap-1a and Sap-1b. It has since been used, with different proteins as bait, to isolate many other cofactors. For example, the progesterone receptor was used as bait to isolate the cofactor SRC-1. The technique has also been employed to map domains that are responsible for protein–protein interactions, by mutagenizing the bait protein. In order for the approach to be successful, it is important that the interacting proteins do not activate transcription when expressed alone in yeast. In many instances the bait is expressed as a fusion with a heterologous DNA-binding domain, in which case it is referred to as a two-hybrid screen.

Box 11.1 (contd.)

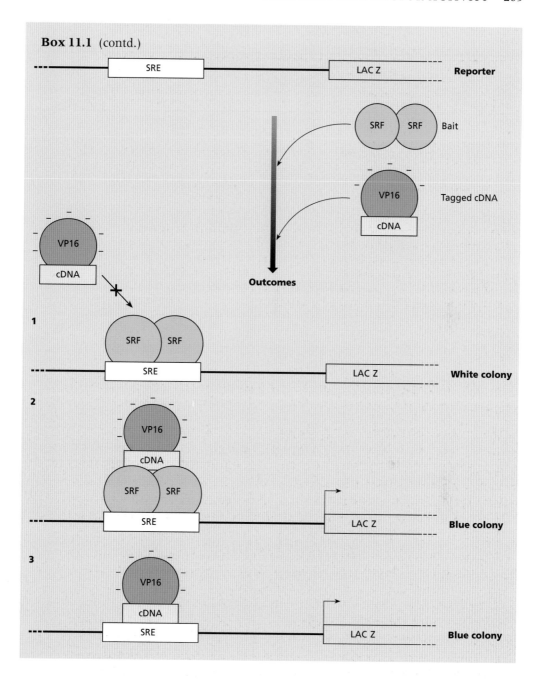

two independent MAPK pathways are primarily involved in transducing stress and cytokine stimuli to the nucleus. One of these pathways results in the activation of **J**un **N**-terminal **k**inases (JNKs; also called **s**tress-**a**ctivated **p**rotein **k**inases or SAPKs), whereas the other switches on a kinase called p38. The JNKs are stimulated rapidly and potently following UV irradiation and are also

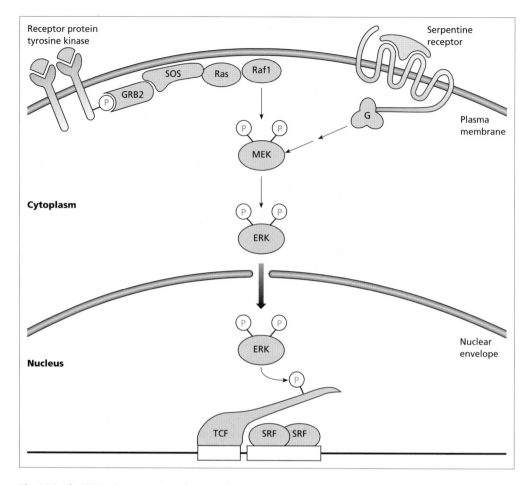

Fig. 11.3 The ERK mitogen-activated protein kinase (MAPK) pathway. Following growth factor stimulation, receptor protein tyrosine kinases activate Ras via the GRB2/SOS complex at the plasma membrane. GTP-bound Ras binds and activates Raf family members, causing them to phosphorylate the cytoplasmic kinases MEK1 and MEK2. These, in turn, phosphorylate and activate ERK1 and ERK2, a fraction of which enter the nucleus and stimulate the transcription of serum-responsive genes by phosphorylating and activating TCF proteins. G protein-coupled serpentine receptors also link to this pathway.

switched on in cells expressing some oncogenic mutants of Ras; as such, they may play an important role in transducing tumour-promoting signals into the transcriptional activation of oncogenes such as *c-fos*. The p38 MAPK is activated in response to the pro-inflammatory cytokine interleukin 1 (IL-1), bacterial lipopolysaccharide and osmotic stress.

Individual members of the TCF family, such as Elk-1 and Sap-1a, show differential responses to the three MAPK pathways (Fig. 11.4). Elk-1 is an efficient substrate for ERKs, JNKs and p38, and is therefore likely to be a point of convergence for the various MAPK cascades. By contrast, Sap-1a is activated preferentially by ERK and p38. IL-1 triggers activation of both Elk-1

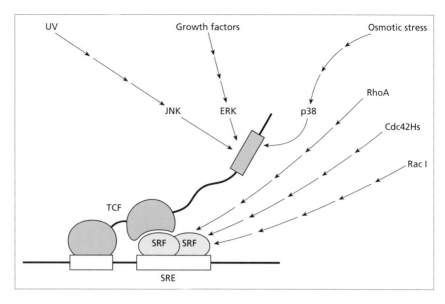

Fig. 11.4 Signalling pathways converge on the *c-fos* SRE. Three distinct MAPK signalling cascades converge on TCFs that associate with SRF at the SRE. The ETS domain of TCFs binds to a DNA sequence to one side of the SRE. This allows an adjacent domain to contact DNA-bound SRF. Transcriptional activation occurs following the phosphorylation of TCFs at sites near their C-termini by ERK, JNK and p38 MAPKs. The TCF Elk-1 is targetted by all three MAPK cascades, whereas the TCF Sap-1a responds efficiently to ERKs and p38, but not JNK. Activation of Elk-1 by ERK-2 requires phosphorylation at seven different sites. SRF can be activated in response to signals emanating from the small GTPases RhoA, Cdc42Hs and Rac1.

and Sap-1a through their phosphorylation by JNK and p38. However, the regulation of TCFs can depend on the cell type and stimulus. For example, in BAC-1 macrophages Elk-1 and Sap-1a are targetted by distinct pathways; Elk-1 is phosphorylated by ERK as part of the Ras pathway, whereas Sap-1a is activated by p38 in a Ras-independent manner. Distinct substrate specificities and the differential use of signalling pathways allow cell type-dependent responses to extracellular signals.

Mutations at the SRE that prevent TCF recruitment reduce substantially the serum induction of the *c-fos* promoter. Nevertheless, SRF itself is able to respond in an ERK- and TCF-independent fashion to lysophosphatidic acid and aluminium fluoride, agents that act through heterotrimeric G proteins. The pathways that regulate SRF have been shown to involve the small GTPases RhoA, Cdc42Hs and Rac1, which play dynamic roles in controlling the actin cytoskeleton. It therefore makes sense that the *β-actin* gene is linked to SREs. Cdc42Hs directs the formation of actin-containing microspikes called filopodia, whereas Rac is required for membrane ruffling and Rho is necessary for the assembly of stress fibres. Chronic activation of Cdc42Hs, Rac and Rho induces morphological changes and malignant transformation. Thus, the combined presence of SRF and TCF allows genes controlled by SREs to respond to a range of signals that are transduced by multiple pathways. This may allow synergistic and coordinate regulation in response to a wide variety of extra-cellular stimuli.

The JNK branch of the MAPK family is also involved in regulating c-Jun (hence the name, **J**un **N**-terminal **k**inase). In order to stimulate transcription, c-Jun must be phosphorylated by JNK at Ser-63 and Ser-73 in its activation domain. In addition to its activation function, which is switched on by phosphorylation, the ability of c-Jun to bind DNA is controlled by phosphorylation in a negative fashion. Thr-231, Ser-243 and Ser-249 are phosphorylated in unstimulated cells; these residues lie immediately adjacent to the bZIP domain and their phosphorylation prevents interaction with DNA, possibly due to electrostatic repulsion. Glycogen synthase kinase 3 (GSK3) and casein kinase II (CKII) can phosphorylate these sites *in vitro* and inhibit DNA binding. It is unclear which kinase(s) performs this function *in vivo*, although AP-1 activity is increased substantially by microinjecting cells with peptides that inhibit CKII. Conversion of Ser-243 to a non-phosphorylatable phenylalanine results in the loss of phosphorylation at all three residues and a considerable activation of c-Jun *in vivo*. Avian sarcoma virus 17 encodes an oncogenic derivative called v-Jun that carries the same substitution, which may release it from negative cellular controls. By contrast, c-Jun appears to exist in a cryptic, inactive form in unstimulated cells. Treatment with the tumour-promoting phorbol ester TPA triggers a cascade of events involving protein kinase C that culminates in the removal of the inhibitory phosphates (Fig. 11.5). Expression of several transforming oncogenes, including *v-Ha-ras*, *v-sis* and *v-src*, can produce a similar response. Indeed, the use of dominant-negative Jun mutants and

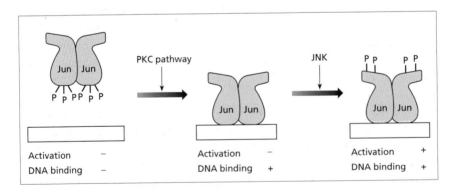

Fig. 11.5 Regulation of c-Jun by phosphorylation. C-Jun is phosphorylated at five major sites. Two residues within the N-terminal activation domain are phosphorylated by JNK in response to various cell stresses. Three residues adjacent to the DNA-binding domain can be phosphorylated by GSK-3 or CKII in order to inhibit DNA binding. Activation of the PKC pathway by tumour-promoting phorbol esters leads to dephosphorylation of the C-terminal residues, thereby allowing c-Jun to bind to its target DNA sites. However, it only stimulates transcription after JNK has phosphorylated the two sites in the activation domain.

c-jun knockout mice has shown that functional c-Jun is essential for transformation by Ha-Ras and v-Src.

It is not necessary to describe here any further instances of transcriptional regulation through phosphorylation, as examples crop up throughout this book. As explained in Chapter 4, TFIIH causes pol II to initiate transcription by phosphorylating its C-terminal domain (CTD). Chapter 5 described how the cAMP-dependent protein kinase A (PKA) activates CREB by phosphorylating Ser-133, thereby allowing it to recruit the cofactors CBP and p300. The previous chapter provided several examples of transcription factors that are transported into the nucleus (NF-κB, STATs), or out of it (NF-AT, Pho4, SWI5), in response to specific phosphorylation events. Phosphorylation can trigger dimerization, as in the case of STATs, or proteolytic degradation, as seen for IκB and β-catenin. Phosphorylation by the cyclin-dependent kinases is an important mechanism for coordinating gene expression with the phases of the cell cycle, as will be described in Chapter 13. Table 11.1 provides an extensive list of instances in which phosphorylation is used as a mechanism for controlling transcription.

Regulation by dimerization

Several classes of transcription factor can only function as dimers. Examples include the STATs, bZIP proteins such as Jun and bHLH proteins such as MyoD. If heterodimerization is involved, the activity of one factor can be strongly regulated via its partner. Clearly, a protein that relies on another for its function will be inactive if its partner is absent. For example, overexpression of MyoD in transfected Cos cells fails to efficiently activate muscle-specific reporter genes unless its dimerization partner E2A is provided in excess. Apart from this obvious dependence, the diversity offered by dimerization provides enormous scope for regulation, with different combinations potentially displaying unique behaviour. DNA-binding properties can be altered dramatically by heterodimerization. For example, Fos–Jun dimers bind to AP-1 sites much more efficiently than Jun homodimers; this reflects differences in dimerization stability, with Fos–Jun combinations forming at least 500-fold more efficiently than Jun–Jun dimers.

As already described, phosphorylation at serines 63 and 73 substantially enhances the ability of c-Jun to activate transcription and this is the basis for its oncogenic cooperation with Ha-Ras. Another member of the Jun family, JunD, contains serine residues at the equivalent positions but is phosphorylated much less efficiently by JNK kinases and is unable to cooperate with Ha-Ras in transforming cells. This is because JunD lacks a JNK docking site, despite having appropriate phosphoacceptor sites. By contrast, c-Jun contains a docking site between residues 30 and 60 that interacts efficiently with a region of JNKs that is adjacent to their catalytic sites; this serves to attract the kinases to their substrate and thereby facilitates phosphorylation of c-Jun. Efficient activation of JunD requires that it dimerizes with c-Jun; the latter can

Table 11.1 Regulation of transcription factor activity by phosphorylation.

Activity regulated	Factor	Detail
Transcriptional initiation	Pol II	CTD phosphorylation by TFIIH triggers transcriptional initiation by pol II
Transcription activation	RB	Phosphorylation by Cdks prevents binding to MyoD and transcription activation
	CREB	Phosphorylation of Ser133 by PKA allows recruitment of CBP and transcription activation
	TCFs	Transcription activation requires phosphorylation near C-terminus by MAPKs
	c-Jun	Phosphorylation of Ser63 and Ser73 by JNK allows transcription activation
DNA binding	c-Jun	Phosphorylation of Thr231, Ser243 or Ser249 by GSK-3 or CKII inhibits DNA binding
	Oct-1	Phosphorylation of Ser385 during mitosis inhibits DNA binding
	E2F	Phosphorylation by cdk2 in S phase prevents DNA binding
Transcription repression	TFIIIB	Phosphorylation by Cdc2 and other mitotic kinases inactivates TFIIIB
	RB	Phosphorylation by Cdks blocks binding and repression of E2F and TFIIIB
	Eve	Phosphorylation of Eve by GSK-3 prevents TBP binding and transcriptional repression
Nuclear transport	Pho4	Phosphorylation of Ser114, Ser128 and Ser152 by Pho85 prevents nuclear accumulation
	SWI5	Nuclear import inhibited after phosphorylation of Ser552, Ser646 and Ser664 by Cdc28
	NF-AT	Dephosphorylation by calcineurin allows nuclear entry
Degradation	IκB	Phosphorylation by IKK triggers ubiquitination and proteolytic degradation
	β-catenin	Phosphorylation by GSK3 triggers ubiquitination and proteolytic degradation
Dimerization	STATs	Phosphorylation by JAKs allows dimerization, nuclear entry and DNA binding

then recruit JNKs to the dimer complex, where they can phosphorylate both c-Jun and JunD. In this way, heterodimerization confers on JunD an important property that is not displayed by JunD homodimers (Fig. 11.6). The failure of JunD in cooperative transformation assays probably reflects the fact that most of the c-Jun within a cell is taken up preferentially in complexes with Fos family members. This example demonstrates that the response of a transcription factor to signalling pathways can be strongly influenced by its

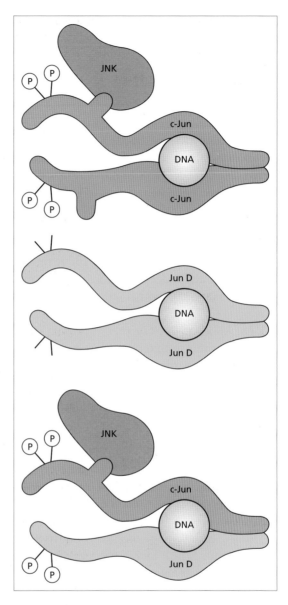

Fig. 11.6 Models of the phosphorylation of c-Jun and JunD by JNKs. C-Jun is phosphorylated and activated very efficiently by JNKs in response to environmental stresses such as UV irradiation. JNKs bind to a docking site between residues 30 and 60 of c-Jun and this brings them close to the phosphoacceptor sites Ser-63 and Ser-73. JunD has comparable phosphoacceptor sites but lacks a docking site that would allow efficient recruitment of JNKs. JunD can, however, be phosphorylated and activated by JNKs if it heterodimerizes with c-Jun, because the latter allows interaction with JNK. Mutations in the bZIP dimerization domain of JunD that increase its affinity for c-Jun, also enhance its ability to respond to JNK activation.

dimerization partners. This feature will enhance considerably the regulatory diversity of signal transduction cascades.

A number of cases have been described in which dimerization is used as a mechanism for silencing a particular factor. One of the best examples involves the bHLH factor MyoD and its relatives myogenin, Myf-5 and MRF-4. These proteins play pivotal roles in promoting skeletal muscle development. For example, MyoD is capable of inducing a programme of myogenic differentiation when transfected into certain fibroblasts. It stimulates transcription of many muscle-specific genes, such as the **m**uscle **c**reatine **k**inase (MCK) gene. A family of small proteins called Id (**I**nhibitor of **d**ifferentiation) contain HLH motifs, but lack the basic sequence that precedes it in most bHLH domains. As was described in Chapter 3, the HLH region mediates dimerization, whereas the basic region is responsible for contacting DNA. As a consequence, heterodimers that form between Id and bHLH proteins such as E2A or MyoD are unable to bind DNA and activate gene expression. Thus, transcription of a reporter driven by the MCK promoter is stimulated by transfection of a MyoD expression vector, but this effect is blocked if an Id vector is cotransfected. The abundance of Id proteins falls during muscle cell differentiation, allowing MyoD and its relatives to induce MCK and other myogenic genes. However, forced overexpression of Id can block the muscle differentiation programme.

Other bHLH proteins have been implicated in the development of neurones or blood cells and in some cases these can also be antagonized by members of the Id family. Indeed, Id proteins serve to prevent premature cell cycle withdrawal and differentiation in several important contexts. Targetted disruption of Id2 or Id3 yields mice with haematopoietic defects. When Id1 and Id3 are knocked out together, neuronal precursor cells differentiate prematurely leading to abnormal neurogenesis. These double knockout mice also display vascular malformations in the forebrain. Furthermore, adult mice with a reduced Id dosage will not support the growth of tumour xenografts due to insufficient angiogenesis. Thus the Id proteins are required to maintain the timing of neuronal differentiation and invasiveness of the vasculature. Because the Id genes are expressed at very low levels in adults, they may make attractive targets for anti-angiogenic drug design.

Another factor that can inhibit gene expression through heterodimerization is C/EBPζ. This bZIP protein interacts avidly with other members of the C/EBP family, but the resultant heterodimers have reduced affinity for specific recognition motifs owing to an unusual residue in the DNA-binding domain of C/EBPζ. As a consequence, C/EBPζ acts as a dominant negative regulator of genes with C/EBP-binding sites and it antagonizes the ability of C/EBPα and C/EBPβ to promote adipogenesis. Production of C/EBPζ is induced by a range of cellular stresses, such as toxins, metabolic inhibitors and nutrient deprivation.

These examples illustrate how dimerization can be used as a regulatory mechanism and how competition between family members may be extremely important. If a factor can be activated or repressed depending on its choice of partners, then transcriptional activity will be dictated by the balance of these proteins within a cell and their relative strengths of interaction. Understanding how a particular gene is controlled in a given tissue can therefore be a daunting task; besides identifying the essential DNA sequences in promoters and enhancers, it may be necessary to determine all the proteins that have the potential to bind to these sites, measure their abundance and distribution in the cell, detail how they can interact and then ascertain the functional consequences of such interactions.

Regulation by proteolysis

Proteolytic degradation is an obvious and drastic way to regulate a transcription factor irreversibly. An example of this form of control, which was encountered in Chapter 10, is the proteolysis of IκB as part of the pathway that leads to NF-κB activation. The processing of p105 into a form that can enter the nucleus and control gene expression was also described. In both these cases, the substrate protein is tagged for proteolysis by the covalent attachment of ubiquitin molecules. Another important example of control by ubiquitination and degradation is provided by p53.

As mentioned in Chapter 3, p53 is an important tumour suppressor protein that is highly conserved among vertebrate species. It displays a variety of bio-chemical activities, including the ability to regulate transcription. It can bind to DNA in a sequence-specific fashion and stimulate the expression of class II genes. Transcriptional activation is mediated through an acidic domain at the N-terminus that binds directly to TATA-binding protein (TBP) and TBP-associated factors (TAFs) in the TFIID complex. In addition to activating genes with p53-binding sites, p53 can also function as a specific repressor of some pol II and pol III promoters that lack its response element. For example, p53 has been shown to inhibit specifically the synthesis of c-fos, cyclin A, tRNA and 5S rRNA. Because these products are involved in promoting growth and cell cycle progression, the transcriptional repression function of p53 may con-tribute to its ability to suppress proliferation and/or tumour formation.

p53 plays a pivotal role in protecting higher organisms against DNA damage. Most normal cells contain very low levels of p53 protein, but it accu-mulates rapidly if they are exposed to genotoxic stresses such as radiation, hypoxia or chemotherapeutic drugs. This increase occurs primarily through a post-translational mechanism, involving a slowing down of p53 turnover. For example, UV irradiation can increase the half-life of p53 from ~30 minutes to ~150 minutes. As a consequence of these changes, the transcription of certain p53-responsive genes is enhanced dramatically. This can result in two strik-ingly different types of cellular response, arrest and apoptosis. In many cases, p53 triggers cell cycle arrest, often in G1 phase. This is thought to be achieved, at least in part, by the induction of a gene called $p21^{Cip1/Waf1}$. The p21 product of this gene binds and inactivates several of the cyclin-dependent protein kinases that are required to drive cell cycle progression; as a consequence, the arrested cells have time to repair their damaged DNA before it gets replicated. There is also evidence that p53 can interact directly with the DNA replication and repair machineries, inhibiting the former and stimulating the latter. These effects are extremely important, because they prevent the propagation of genetic mutations which might result in diseases such as cancer. A far more

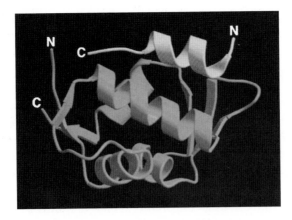

Fig. 11.7 Cocrystal structure showing a complex involving the N-terminal domains of p53 and Mdm2. The X-ray crystal structure of the 109 N-terminal residues of Mdm2 is binding a 15-residue peptide from the N-terminal transcription activation domain of p53. The p53 peptide forms an amphipathic α-helix that is buried in a deep hydrophobic cleft within Mdm2. The interface relies on precise steric complementarity between the cleft and the hydrophobic face of the p53 α-helix, especially residues Phe[19], Trp[23] and Leu[26]. These same three residues are crucial for p53 to contact TFIID and activate transcription. Binding to Mdm2 therefore blocks the ability of p53 to stimulate gene expression. Reproduced with permission from Kussie *et al.*, Structure of MDM2 oncoprotein bound to the p53 tumour suppressor transactivation domain. *Science* **274**. Copyright 1996, American Association for the Advancement of Sciences.

drastic mechanism for eliminating the damaged DNA is for p53 to induce apoptosis (programmed cell death). This is an important protective mechanism for removing cells with DNA damage from sunburnt skin. Furthermore, the effectiveness of many cancer therapies is thought to stem from their ability to induce a p53-dependent apoptotic response. Several genes that are involved in apoptosis have been shown to be induced by p53, including *bax*, which is sometimes referred to as a 'death gene'. Because of its central role in ridding the body of mutated DNA, p53 has been given the rather colourful title 'guardian of the genome'.

Another target for p53 activation is an oncogene called *Mdm2*, which encodes a transcription factor. The Mdm2 protein contains a deep hydrophobic cleft that swallows the amphipathic α-helical activation domain of p53, which is responsible for binding TFIID and stimulating pol II transcription (Fig. 11.7). With its activation domain masked in this way, p53 loses the capacity to regulate its target genes. Furthermore, interaction with Mdm2 destabilizes p53 and promotes its rapid proteolytic degradation. The Mdm2-binding domain of p53 is sufficient to target heterologous proteins for rapid destruction. These proteins therefore form an autoregulatory feedback loop: genotoxic stresses cause a rise in p53 levels, which stimulates the synthesis of Mdm2; Mdm2 then binds p53, blocks its activation function and promotes its proteolytic turnover, thereby returning the cell to its pre-stressed level of p53. DNA repair must occur during the interval in which p53 levels are elevated. A clear indication of the importance of Mdm2 in controlling the abundance of p53 has been provided by gene disruption studies in mice; specific knockout of the *Mdm2* gene is lethal unless the *p53* gene is also deleted, because *Mdm2* is essential to prevent the deregulation of *p53*.

The p53 gene is lost or mutated in ~55% of all solid human tumours, the highest frequency of any known gene. In most of these cases the mutations arise somatically, but germ-line p53 mutations are transmitted in cancer-prone families with a condition called Li–Fraumeni syndrome. The majority of mutations that arise in p53 are found at certain 'hot-spot' sites within its DNA-binding domain, thereby preventing it from activating transcription. In some of the tumours that retain wild-type p53, its function is found to be severely compromised due to the amplification and overexpression of Mdm2. Although much less frequent than p53 disruption, abnormally elevated Mdm2 levels are found in ~30% of soft tissue sarcomas and are often associated with a poor prognosis. Once a tumour has rid itself of p53 function, it is released from a principle surveillance mechanism and is able to accumulate further mutations much more freely, often resulting in malignant transformation. Inactivation of p53 is therefore likely to be a very important step in carcinogenesis. This is clearly illustrated by the fact that *p53^{-/-}* mice develop normally, but display a strong predisposition to cancer, such that 74% develop tumours by the age of 6 months. Many current gene therapy schemes aim to restore p53 function to malignant cells. For example, clinical trials have involved injecting lung cancer tissue with a retroviral vector carrying the p53 gene driven from a constitutively active β-actin promoter; this was found to induce local apoptosis frequently and sometimes resulted in measurable tumour regression. Other schemes aim to inactivate endogenous Mdm2 or to block the ability of Mdm2 to associate with p53.

Because of its importance as a defence mechanism, p53 is targetted by several DNA tumour viruses. The most clinically important of these are the **human papillomaviruses** (HPVs), which contribute to the development of nearly all cervical cancers, as well as vulval, penile and perianal cancers. One of the viral products is an oncoprotein called E6, which binds to p53 and targets it for rapid degradation through ubiquitin-directed proteolysis. As a consequence, HPV-infected cells are unable to accumulate high levels of p53 in response to genotoxic stresses; they display marked genetic instability and cannot protect themselves against oncogenic mutations. Because E6 expression is functionally equivalent to alterations within p53 itself, cervical cancers differ from most major epithelial malignancies in showing a very low incidence of somatic mutation in p53.

Poliovirus utilizes proteolytic degradation to shut down the expression of host cell genes. Viral infection results in the rapid inhibition of host cell RNA synthesis by all three classes of nuclear RNA polymerase. Pol I transcription is downregulated most rapidly, within 2–3 hours of infection, followed by pol II after 3–4 hours and pol III at 4–5 hours postinfection. In each case, regulation has been correlated with the inactivation of specific transcription factors. For the class I system, transcriptional silencing appears to be due to the inactivation of pol I and/or its associated factors. The shut-off of pol II transcription correlates with a specific decrease in the activity of TFIID. This effect can be reproduced by incubating TFIID fractions with a poliovirus-encoded protease called 3C. Recombinant TBP can be cleaved near its N-terminus by protease 3C *in vitro* and the proteolysed form of TBP is detected *in vivo* after 4 hours of infection. Because TBP cleavage is first observed 1–2 hours after the shut-off of rRNA synthesis, it is unlikely to be responsible for pol I inhibition. However, the timing of this event is consistent with a role in regulating pols II and III. This has yet to be demonstrated, and other components of the TFIID complex could be the critical targets. In addition, although TFIIIB activity is reduced following poliovirus infection, the greatest effect on the class III transcription apparatus is due to the regulation of TFIIIC2. The gene encoding protease 3C is sufficient to inactivate TFIIIC2 and class III gene expression when transfected into HeLa cells. At least two subunits of TFIIIC2 are cleaved by protease 3C *in vitro*, generating a factor that is transcriptionally inactive, although still able to bind DNA. Protease 3C is one of the two poliovirus-encoded proteinases that are involved in processing the viral polyprotein precursor into capsid and non-capsid proteins. These enzymes are thought to be highly specific in their action, as two-dimensional gel comparisons of infected and mock-infected extracts revealed fewer than 10 polypeptides that had been cleaved in response to poliovirus. It is therefore very striking that several components of the basal transcription apparatus are amongst the targets.

Regulation by ligand binding

The nuclear receptors provide a direct signal transduction system, in which the same polypeptide serves to receive a stimulus from outside the cell and convert this into a transcriptional response. This superfamily of ligand-dependent transcription factors plays crucial roles in controlling reproduction, development and tissue homeostasis. It includes receptors for steroid hormones, thyroid hormones, peroxisomal activators and the hormonal forms of vitamins A and D (Fig. 11.8). Because these ligands are relatively small and hydrophobic, they can enter a target cell by simple diffusion. This is a slow process, in molecular terms, and so is generally used to control gradual changes. It also lacks the advantage of signal amplification that can be achieved using kinase cascades.

The two best characterized groups of nuclear receptors are the steroid hormone receptors and the RXR heterodimers. The steroid receptors include the oestrogen receptor (ER), glucocorticoid receptor (GR) and the progesterone receptor (PR). The second group consists of factors such as the thyroid hormone receptor (TR) and the all-*trans* retinoic acid receptor (RAR), which can only bind DNA with high affinity after heterodimerizing with the retinoid

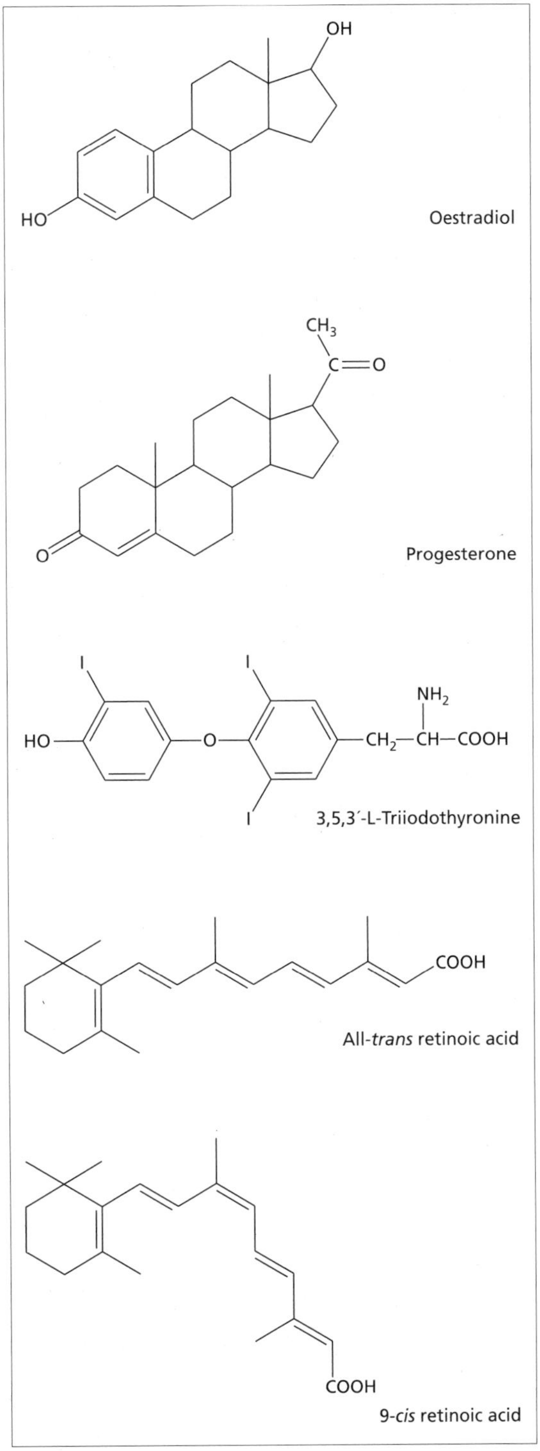

Fig. 11.8 Structures of some of the ligands for nuclear hormone receptors. Oestradiol is the ligand for ER. Progesterone is the ligand for PR. 3,5,3′-L-triiodothyronine is the ligand for TR. All-*trans* retinoic acid is the ligand for RAR and 9-*cis* retinoic acid is the ligand for RXR.

Fig. 11.9 Nuclear receptor organization. A typical nuclear receptor contains an N-terminal region with a transcription activation function (AF-1). The size of this region can vary substantially between different receptors. It is followed by the DNA-binding domain (DBD), which is highly conserved between family members. The DBD is connected by a variable hinge region to the ligand-binding domain (LBD). As well as hormone recognition, this region allows dimerization and contains a transcription activation function (AF-2).

X receptor (RXR). RXR does not need to bind a ligand in order to function in such heterodimers. However, it can also function as a ligand-dependent homodimer which binds to 9-*cis* retinoic acid.

All members of the nuclear receptor superfamily share a similar basic organization (Fig. 11.9). The most conserved region is the DNA-binding domain, which is a compact, globular structure built around two zinc ions that are both coordinated tetrahedrally to four cysteine residues. As was described in Chapter 3, the more N-terminal of these zinc ions is flanked by an α-helix which makes base-specific contacts in the major groove of DNA. This recognition helix can identify a hexameric 'half-site' of the hormone response element. High-affinity binding depends on homo- or heterodimerization, such that each monomer contacts one hexamer of the DNA site. The major dimerization interface is located within a C-terminal domain of approximately 220 residues. This region is joined to the DNA-binding domain by a poorly conserved hinge region. The C-terminal domain is also responsible for binding to the ligand and contains a transcription activation function, which is often referred to as AF-2 (**a**ctivation **f**unction-2). An additional activation function (AF-1) can be found in a poorly conserved region of variable length that is located N-terminal to the DNA-binding domain; this function is not controlled directly by the ligand, but cooperates with the C-terminal activation function in a cell- and promoter-specific manner.

The ligand-binding domain effectively serves as a molecular switch that can shift nuclear receptors between active and inactive states. Chapter 8 has already described how the TR and RAR can be converted by the presence of ligand from a dominant silencer, that is associated with histone deacetylase corepressors, to a transcriptional activator that interacts with multiple coactivators with **h**istone **a**cetyltransferase (HAT) activity. In addition to CBP, p300 and p/CAF, which associate with many transcription activators (e.g. CREB, AP-1, STATs), a bewildering array of coactivators have been identified that appear to be more specific to the nuclear receptor superfamily. An example is SRC-1 (**s**teroid **r**eceptor **c**oactivator-1), which was isolated using the ligand-bound progesterone receptor as bait in a yeast two-hybrid screen. Another example is ACTR (**ac**tivator of the **T**R and **R**AR), which was isolated indendently by several approaches; the most interesting involved the microdissection of chromosomal regions which become amplified in breast cancer—hence, an alternative name is **a**mplified **in** **b**reast cancer (AIB)1. The fact that ACTR/AIB1 is overexpressed due to amplification in primary breast tumours, as well as several ER-positive breast and ovarian cancer cell lines, suggests that it may contribute to the development of steroid-dependent cancers. ACTR, SRC-1, and a third coactivator called SRC-2, are all related, with 30–35% amino acid identity. Each contains a basic helix-loop-helix (bHLH) motif, HAT activity and two major transactivation domains. Like p/CAF, CBP and p300, they interact with nuclear hormone receptors in an agonist-dependent manner. This association requires the AF-2 activation region within the **l**igand-**b**inding **d**omain (LBD). In addition to all these HATs, AF-2 can also recruit in a ligand-dependent manner a coactivator called DRIP or TRAP. The DRIP/TRAP complex contains 14–16 subunits and resembles the mediator complex which associates with the **C**-terminal **d**omain (CTD) of pol II; it may therefore serve to target the pol II holoenzyme to hormone-responsive promoters.

The AF-2 sequence is located towards the C-terminus of the LBD and is well conserved between receptors. Indeed, crystallographic analysis has shown that the overall structure of the LBD is very similar between the TR, RAR, RXR and ER; in each case it involves 11 or 12 highly conserved α-helices. The ligand becomes buried deep within the hydrophobic core of a binding pocket inside the LBD. The AF-2 of the ER is centred on helix 12, which forms a lid for the ligand-binding cavity. This lid closes when oestradiol binds, allowing helix 12 to present a series of charged residues away from the cavity and towards any incoming coactivators. However, the AF-2 undergoes extensive rearrangements when the ER binds the antagonist raloxifene, which is used clinically to counter the mitogenic effects of oestradiol. Raloxifene is too bulky to allow the lid to close, so it causes helix 12 to rotate by 130° and shift by 10° with respect to its position when oestradiol is bound; as a consequence, it lies in a groove which partially buries a conserved lysine that is required to activate transcription (Fig. 11.10). This can explain why raloxifene prevents the recruitment of agonist-dependent coactivators. These data may provide the basis for the structure-based design of improved antagonists for the treatment of oestrogen-associated diseases.

Peroxisome **p**roliferator-**a**ctivated **r**eceptor **g**amma (PPARγ) is a nuclear receptor that dimerizes with RXR and plays a key role in adipocyte differentiation. It is the target of thiazolidinedione drugs, which are used to treat type 2 diabetes by improving the body's sensitivity to insulin. Rare cases of combined hypertension, severe insulin resistance and diabetes mellitus have arisen from substitutions in the LBD of PPARγ. These mutations reduce both the ligand-binding and coactivator-binding activity of the receptor; its ability to stimulate transcription is severely compromised as a consequence. One kindred was found to have a substitution in helix 12 of the LBD, whereas another patient had a mutation

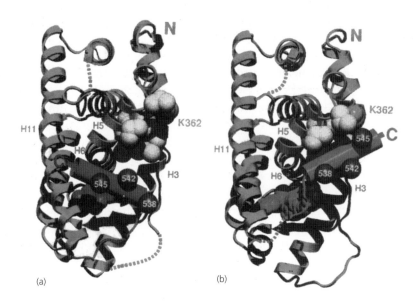

Fig. 11.10 Crystal structures of the ER LBD bound to the agonist oestradiol (a) or the antagonist raloxifene (b). It is essential for the AF-2 activation function and its mutation can result in a receptor that is unresponsive to hormone. In the presence of oestradiol, helix 12 sits over the ligand-binding cavity, with its hydrophobic surface facing in towards the steroid and its charged face exposed on the surface where it can bind to coactivators. By contrast, the antagonist raloxifene prevents helix 12 from sitting over the cavity; its position is rotated by 130° relative to the agonist structure and this buries a conserved lysine that is necessary for coactivator recruitment. It therefore seems that the antagonistic properties of raloxifene result from its ability to prevent AF-2 adopting a transcriptionally competent conformation. Adapted with permission from *Nature* **389**. Copyright Macmillan Magazines Limited.

which affects the surface that packs against helix 12. In both cases the orientation of helix 12 is likely to be perturbed, thereby disrupting AF-2 and the ligand-dependent transcriptional response. Several cases of obesity have been linked to substitutions near the N-terminus of PPARγ, which impair its phosphorylation by MAPK; these missense mutations have a gain-of-function effect that enhances adipocyte differentiation. The treatment of human metabolic disorders, such as type 2 diabetes, hypertension and obesity may benefit considerably from the development of more potent and specific ligands for PPARγ.

As described in Chapter 8, some members of the nuclear hormone receptor superfamily function as transcriptional silencers in the absence of an appropriate agonist. This repression function resides in the LBD, but is separable from AF-2. It interacts with the related corepressors N-CoR (**n**uclear receptor **cor**epressor) and SMRT (silencing mediator for RXR and TR). Genes encoding these cofactors were isolated using two-hybrid screens in yeast. As well as TR, RAR and RXR, N-CoR binds to ER in the presence of the antagonist tamoxifen and PR in the presence of the antagonist RU-486. This may explain how antagonist-bound steroid receptors can actively silence transcription.

The RAR–RXR heterodimer binds with high efficiency to a recognition sequence that is organized as a direct repeat with either a 5-bp spacing (DR5) or a single nucleotide spacing (DR1). On DR5 elements, RXR binds upstream and RAR binds downstream; retinoic acid induces release of N-CoR, recruitment of coactivators and transcriptional activation. By contrast, the polarity of the complex is reversed on DR1 elements, with RAR upstream and RXR downstream. In such cases, N-CoR remains bound to RAR even in the presence of retinoic acid and transcription remains silenced despite the recruitment of positive cofactors. Thus, N-CoR is the dominant influence in this case, and DR1 motifs do not confer responsiveness to all-*trans* retinoic acid (Fig. 11.11). However, RXR can also bind to DR1 sites as a homodimer and in this situation it is hormone responsive. In effect, RAR serves as a repressor of RXR at DR1 sites. Thus, a change in the spacing of a DNA recognition site allows alternative responses to ligand availability. The subtlety of this control is likely to be invaluable during the complex developmental programmes that are coordinated by retinoids.

Whereas RAR–RXR and TR–RXR bind tightly to their target DNA sites in the presence or absence of ligand, subcellular fractionation has shown that a substantial proportion of unliganded steroid receptors can be found in the cytoplasm. This may be due, at least in part, to artifactual leakage of receptors from the nucleus during experimental manipulation. Nevertheless, an important body of opinion considers that the cytoplasmic presence of hormone-free receptors is functionally significant. Less controversial is the evidence that steroid receptors associate, when ligand is unavailable, with a large multi-component complex of molecular chaperones. This complex includes Hsp90 and the immunophilin Hsp56, and serves to maintain unliganded receptors in an inactive but hormone-accessible 'poised' conformation, ready to respond to signal. Interaction of steroid receptors with Hsp90 appears necessary for efficient ligand binding and regulation; it is presumed that this reflects the ability of chaperones to influence molecular folding. Agonists are thought to trigger dissociation of the Hsp90 complex, allowing receptor dimerization, nuclear translocation, DNA binding and transcriptional activation (Fig. 11.12). Certain mutants that lack the region which interacts with hormone and Hsp90 are constitutively active, which suggests that this signal-response domain may also maintain ligand-free receptors in a transcriptionally inactive state. Indeed, fusion of the Myc oncoprotein to the ER LBD confers hormone dependence upon Myc activity, allowing cell transformation in an oestrogen-dependent manner.

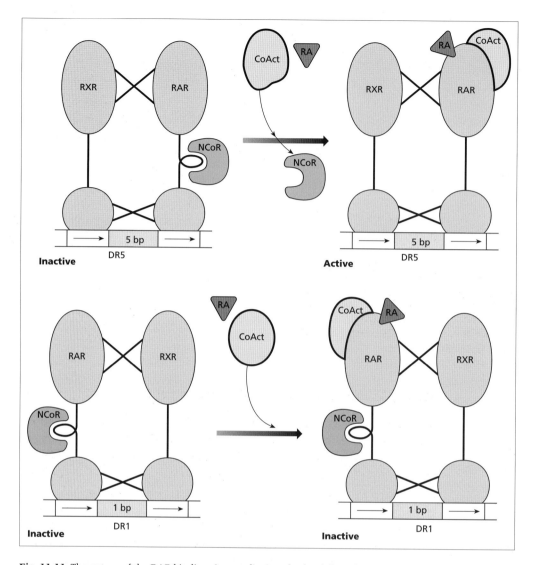

Fig. 11.11 The nature of the RAR binding site can dictate whether it functions as a hormone-dependent activator. A heterodimer composed of the retinoic acid receptor (RAR) and the retinoid X receptor (RXR) will bind to DNA recognition sequences in which the half-sites are separated by either 5 bp (DR5) or 1 bp (DR1). At a DR5 site, RAR is positioned downstream of the RXR; addition of retinoic acid (RA) triggers the release of N-CoR and the recruitment of coactivators, resulting in transcriptional stimulation. When bound to a DR1 site, the RAR is located upstream of RXR. In this configuration, RA is unable to dismiss N-CoR from the hinge region connecting the DNA-binding and ligand-binding domains; although the coactivator complex is still recruited in response to RA, the repressive effect of N-CoR is dominant and transcription remains silenced.

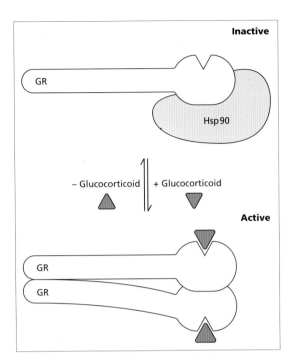

Fig. 11.12 A complex of molecular chaperones including Hsp90 binds to steroid receptors in the absence of hormone but not in its presence. Unliganded steroid receptors (e.g. glucocorticoid receptor, GR) are found in transcriptionally inactive complexes containing Hsp90 and other molecular chaperones. Such complexes are not bound to DNA and may often be found in the cytoplasm. Binding of an agonist, in this case glucocorticoid, triggers release of the Hsp90 complex, allows dimerization and converts the receptor into a transcriptionally active form.

Multiple mechanisms to regulate an individual factor

An individual transcription factor can often be controlled by several different regulatory mechanisms. This increases the flexibility with which it can respond to environmental stimuli, integrating the information provided by multiple signals. For example, the availability of nuclear hormone receptors is dictated in the first instance by the tissue-specific transcription patterns of the genes that encode them. When present in a cell, the activity of these receptors only becomes manifest in the presence of the appropriate ligand, such as a steroid or retinoic acid. Their function can be further modulated by phosphorylation and is sometimes dependent on the availability of a suitable dimerization partner. The ability of these receptors to regulate transcription is also influenced strongly by the presence of coactivators and corepressors. This degree of regulatory complexity is probably common to a large proportion of factors. Another example is provided by the Rel proteins, which are controlled at the level of transcription, alternative splicing, combinatorial dimerization, subcelluluar compartmentalization, phosphorylation, selective proteolysis and protein–protein interactions. Different mechanisms may operate over different time scales; for example, phosphorylation occurs much more rapidly than changes in the rate of synthesis. There is undoubtedly a hierarchy in the importance of individual control steps for any particular factor, some of which are cardinal, whereas others provide coupling to different regulatory pathways and some serve merely to allow subtle fine tuning.

Further reading

Reviews

Beato, M., Herrlich, P. & Schutz, G. (1995) Steroid hormone receptors: many actors in search of a plot. *Cell* **83**: 851–857.

Calkhoven, C. F. & Ab, G. (1996) Multiple steps in the regulation of transcription-factor level and activity. *Biochem. J.* **317**: 329–342.

Carmeliet, P. (1999) Controlling the cellular brakes. *Nature* **401**: 657–658.

De Cesare, D., Fimia, G. M. & Sassone-Corsi, P. (1999) Signalling routes to CREM and CREB: plasticity in transcriptional activation. *Trends Biochem. Sci.* **24**: 281–285.

Fields, S. & Sternglanz, R. (1994) The two-hybrid system: an assay for protein–protein interactions. *Trends Genet.* **10**: 286–292.

Freedman, L. P. (1999) Increasing the complexity of coactivation in nuclear receptor signalling. *Cell* **97**: 5–8.

Hill, C. S. & Treisman, R. (1995) Transcriptional regulation by extracellular signals: mechanisms and specificity. *Cell* **80**: 199–211.

Hunter, T. & Karin, M. (1992) The regulation of transcription by phosphorylation. *Cell* **70**: 375–387.

Jackson, S. P. (1992) Regulating transcription factor activity by phosphorylation. *Trends Cell Biol.* **2**: 104–108.

Jones, N. (1990) Transcriptional regulation by dimerization: two sides to an incestuous relationship. *Cell* **61**: 9–11.

Ko, L. J. & Prives, C. (1996) p53: puzzle and paradigm. *Genes Dev.* **10**: 1054–1072.

Massari, M. E. & Murre, C. (2000) Helix-loop-helix proteins: regulators of transcription in eucaryotic organisms. *Mol. Cell. Biol.* **20**: 429–440.

Norton, J. D., Deed, R. W., Craggs, G. & Sablitzky, F. (1998) Id helix-loop-helix proteins in cell growth and differentiation. *Trends Cell Biol.* **8**: 58–65.

Oren, M. (1999) Regulation of the p53 tumour suppressor protein. *J. Biol. Chem.* **274**: 36031–36034.

Perlmann, T. & Evans, R. M. (1997) Nuclear receptors in Sicily: all in the famiglia. *Cell* **90**: 391–397.

Pratt, W. B. (1993) The role of heat shock proteins in regulating the function, folding, and trafficking of the glucocorticoid receptor. *J. Biol. Chem.* **268**: 21455–21458.

Torchia, J., Glass, C. & Rosenfeld, M. G. (1998) Co-activators and corepressors in the integration of transcriptional responses. *Curr. Opin. Cell Biol.* **10**: 373–383.

Wasylyk, B., Hagman, J. & Gutierrez-Hartmann, A. (1998) Ets transcription factors: nuclear effectors of the Ras-MAP-kinase signalling pathway. *Trends Biochem. Sci.* **23**: 213–216.

Selected papers

Regulation of transcription through phosphorylation

Hill, C. S., Wynne, J. & Treisman, R. (1995) The Rho family GTPases RhoA, Rac1, and CDC42Hs regulate transcriptional activation by SRF. *Cell* **81**: 1159–1170.

Whitmarsh, A. J., Shore, P., Sharrocks, A. D. & Davis, R. J. (1995) Integration of MAP kinase signal transduction pathways at the serum response element. *Science* **269**: 403–407.

Isolation of TCFs using a yeast genetic screen

Dalton, S. & Treisman, R. (1992) Characterization of SAP-1, a protein recruited by serum response factor to the *c-fos* serum response element. *Cell* **68**: 597–612.

Regulation through dimerization

Benezra, R., Davis, R. L., Lockshon, D., Turner, D. L. & Weintraub, H. (1990) The protein Id: a negative regulator of helix-loop-helix DNA binding proteins. *Cell* **61**: 49–59.

Jen, Y., Weintraub, H. & Benezra, R. (1992) Overexpression of Id protein inhibits the muscle differentiation program: *in vivo* association of Id with E2A proteins. *Genes Dev.* **6**: 1466–1479.

Kallunki, T., Deng, T., Hibi, M. & Karin, M. (1996) c-Jun can recruit JNK to phosphorylate dimerization partners via specific docking interactions. *Cell* **87**: 929–939.

Lyden, D., Young, A. Z., Zagzag, D. *et al.* (1999) Id1 and Id3 are required for neurogenesis, angiogenesis and vascularization of tumour xenografts. *Nature* **401**: 670–677.

p53 and Mdm2

Haupt, Y., Maya, R., Kazaz, A. & Oren, M. (1997) Mdm2 promotes the rapid degradation of p53. *Nature* **387**: 296–299.

Kubbutat, M. H. G., Jones, S. N. & Vousden, K. H. (1997) Regulation of p53 stability by Mdm2. *Nature* **387**: 299–303.

Kussie, P. H., Gorina, S., Marechal, V. *et al.* (1996) Structure of the MDM2 oncoprotein bound to the p53 tumour suppressor transactivation domain. *Science* **274**: 948–953.

Nuclear hormone receptors

Barroso, Gurnell, M., Crowley, V. E. F. *et al.* (1999) Dominant negative mutations in human PPARγ associated with severe insulin resistance, diabetes mellitus and hypertension. *Nature* **402**: 880–883.

Brzozowski, A. M., Pike, A. C. W., Dauter, Z. *et al.* (1997) Molecular basis of agonism and antagonism in the oestrogen receptor. *Nature* **389**: 753–758.

Chen, J. D. & Evans, R. M. (1995) A transcriptional corepressor that interacts with nuclear hormone receptors. *Nature* **377**: 454–457.

Horlein, A. J. *et al.* (1995) Ligand-independent repression by the thyroid hormone receptor mediated by a nuclear receptor corepressor. *Nature* **377**: 397–404.

Nagy, L., Kao, H.-Y., Chakravarti, D. *et al.* (1997) Nuclear receptor repression mediated by a complex containing SMRT, mSin3A, and histone deacetylase. *Cell* **89**: 373–380.

Picard, D., Salser, S. J. & Yamamoto, K. R. (1988) A movable and regulable inactivation function within the steroid binding domain of the glucocorticoid receptor. *Cell* **54**: 1073–1080.

Chapter 12: Cell Cycle Regulation of Transcription

Cell populations that are actively proliferating go through a precisely ordered series of events to ensure that all their components have been duplicated before they divide, a process that is referred to as the cell cycle. Chromosomal replication occurs during a defined interval called S phase (for DNA **S**ynthesis), whereas the period of cell division is referred to as M phase or **m**itosis. In between the S and M phases come two **G**ap phases, called G1 and G2 (Fig. 12.1). These cyclic transitions can have a very profound effect upon gene expression. Perhaps the most dramatic episode is mitosis, which in higher organisms is accompanied by a general repression of transcription. In addition, the expression of a minority of genes fluctuates in a phase-dependent manner during the interphase period between mitoses (G1, S and G2). For example, histone synthesis occurs specifically during S phase, which ensures that supply coincides with demand as chromosomal DNA is replicated. Other examples of cell cycle-regulated genes include those encoding most cyclins and many of the enzymes required for DNA synthesis. The cyclical patterns of regulation are frequently conserved across vast evolutionary distances, which suggests that they are of fundamental importance. Indeed, loss of appropriate cell cycle regulation can result in genomic instability and is often found in tumours. An understanding of cell cycle control can provide major insights into hyperproliferative conditions such as cancer and has suggested ways in which such diseases may be treated.

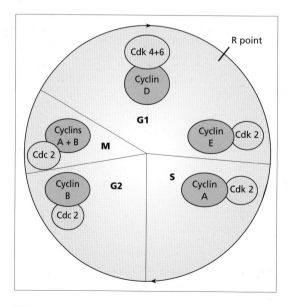

Fig. 12.1 The four phases of the somatic mammalian cell cycle. Cell division occurs during M phase (mitosis). The period between successive mitoses is called interphase and consists of two gap phases, G1 and G2, separated by the S phase when chromosomes are duplicated. The restriction (R) point occurs approximately two thirds of the way through G1, and marks the point at which cells become committed to replicate and divide irrespective of the presence of external growth factors. Also shown are the various cyclins and their associated kinases (cdks) at the stages at which they are thought to act.

Approximately 7% of all class II genes are periodically expressed in the budding yeast *Saccharomyces cerevisiae*. In many of these cases, the periodicity is not important and the genes can be expressed ectopically from a constitutive promoter without any deleterious effects. However, in some cases the periodicity is essential for maintaining the correct order of events. A clear example of this is provided by several of the genes encoding cyclins, which activate the kinases that drive the cell cycle and control the transitions between different phases (Fig. 12.1). The most dramatic periodicity of any gene in *S. cerevisiae* is displayed by *CLN1*, which encodes a cyclin and is induced 25-fold in late G1 phase. In both budding yeast and humans, the activity of the G1-expressed cyclins is critical in deciding whether the cell should commit itself to another round of replication and division, the alternative being to leave the cycle and enter a quiescent resting state (G0 or stationary phase) or a differentiation pathway. This decision is reached in response to various positive and negative extracellular growth signals that control the synthesis and activity of the G1 cyclins. If conditions are propitious, cells accumulate active G1 cyclins which trigger passage into S phase. The moment in late G1 at which cells become committed to enter S phase is called Start in yeast and the restriction point in mammals. Once this stage is passed, the cell cycle machinery becomes engaged, resulting in an ordered progression through chromosome replication and division. Following mitosis, cells again assess their environment and decide whether to continue proliferating or to exit the cycle, according to the levels of the G1 cyclins. A substantial proportion of this fundamental control process is executed through the activity of specific transcription factors.

Transcriptional control at the G1/S transition in yeast

Prior to Start, yeast have the option of entering a developmental pathway involving sexual conjugation and meiotic division. Passage through Start, however, commits them to complete the mitotic cycle. The decision between these alternatives is determined to a large extent by the levels of the G1 cyclins Cln1, Cln2 and Cln3. Constitutive overproduction of these Clns leads to premature entry into the cycle, which suggests that G1 cyclin activity is rate limiting for starting the cell cycle. The *CLN1* and *CLN2* genes are expressed periodically, under the influence of the factor SBF, which was encountered in Chapter 8 in the context of the *HO* promoter. Multiple copies of the SBF recognition motif will confer G1/S-specific transcription on a heterologous reporter gene. SBF contains two polypeptides, Swi4 and Swi6; Swi4 is responsible for DNA binding, whereas Swi6 performs an accessory role. *In vivo* footprinting has shown that SBF binds to the *CLN2* promoter throughout G1 phase, but it only becomes activated once cells reach Start. Activation is mediated by the G1 cyclin Cln3 bound to the cyclin-dependent kinase Cdc28. Unlike Clns 1 and 2, Cln3 is synthesized continuously in growing cells and activates SBF once it has accumulated to a critical level as cells approach Start. However, the *CLN3*

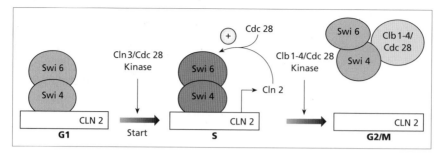

Fig. 12.2 A model for the cell cycle regulation of *CLN2* gene expression. In early G1 phase, the SBF factor containing Swi4 and Swi6 binds to the *CLN2* promoter but is unable to activate transcription. Cells reach Start when their rates of growth and protein synthesis are sufficient to allow Cln3 to accumulate to a critical level. The cyclin-dependent kinase Cdc28 is switched on by Cln3 and activates SBF, thereby stimulating the expression of *CLN2*. The Cln2 product itself interacts with Cdc28, establishing a positive feedback loop that ensures further SBF activation. As yet, it is unclear how Cln-bound Cdc28 activates SBF; mutation of the major phosphorylation sites in Swi4 and Swi6 has no effect on the timing or magnitude of activation, which suggests that Cdc28 may act indirectly on SBF, perhaps by phosphorylating some unidentified coactivator that binds to Swi4 or Swi6. After S phase, the Clns are rapidly degraded. A second wave of cyclins, called Clbs 1–4, associate with Cdc28 during G2 and M phases and these cause SBF to dissociate from the *CLN2* promoter. Phosphorylation of Swi6 at a site adjacent to its nuclear localization signal causes its displacement to the cytoplasm during most of G2 and M phases. At the end of mitosis, Swi6 is dephosphorylated and the Clb cyclins are themselves degraded, allowing SBF to rebind and the cycle to start again.

mRNA contains a short upstream open reading frame that serves as a translational control element which specifically represses Cln3 production when growth and protein synthesis are slow. Passage through Start therefore requires that cells attain a critical rate of protein synthesis, which is achieved once they have reached the size threshold needed for division. In this way, Cln3 production provides a mechanism for coupling S phase entry to translational activity, which correlates closely with the rate of cell growth. If *CLN3* dosage is increased, yeast divide at smaller than normal size with a shortened G1 phase. Under conditions of active growth and protein synthesis, Cln3 accumulates to the threshold level and binds and activates the Cdc28 protein kinase; Cln3/Cdc28 switches on SBF that is bound to the *CLN2* promoter, thereby stimulating transcription (Fig. 12.2). The resultant wave of G1 cyclin activity drives cells into S phase, as well as increasing SBF activity in a positive feedback loop. As a result of this feedback, a small initial increase in SBF activity due to Cln3 can be rapidly amplified; this may contribute to the concerted and irreversible committment that yeast undergo when they pass through Start (Fig. 12.3). In effect, the translational control of *CLN3* provides a mechanism for coupling cell growth with division, which is essential for size homeostasis. At the end of S phase, Clns1–3 are degraded and a set of mitotic cyclins, called Clbs1–4, associate with Cdc28 and cause SBF to dissociate from the *CLN2*

Fig. 12.3 The cyclin cycle in *S. cerevisiae*. The alternating periodic accumulation of G1 cyclins (Clns1–3) and mitotic cyclins (Clbs1–4) is shown relative to the phases of the cell cycles. Once the G1 cyclins reach a threshold level, cells pass through Start; these cyclins are then rapidly degraded and the mitotic cyclins accumulate until they trigger mitosis and are, in turn, degraded, allowing the cycle to begin again.

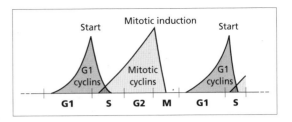

promoter. Swi6 becomes phosphorylated at a consensus Cdc28 site that is adjacent to its nuclear localization signal; as a consequence, it is predominantly cytoplasmic until this residue undergoes dephosphorylation at the end of mitosis.

A factor called MBF is responsible for the S phase-specific activation of a large number of genes encoding proteins required for DNA synthesis, including enzymes responsible for the production of deoxyribonucleotides and components of the DNA replication apparatus. It may also control synthesis of the replication-coupling assembly factor, which facilitates the incorporation of newly replicated DNA into nucleosomes. The periodic expression of these genes may help ensure that replication only occurs at the appropriate time. However, many of these products are very stable and can be expressed constitutively with no adverse effects. MBF is activated by the G1 cyclins, in a manner similar to SBF. MBF is a complex containing the polypeptides Mbp1 and Swi6; it resembles SBF in that they both contain Swi6 and their Mbp1 and Swi4 subunits are closely related, with 50% identity in their DNA-binding domains (Fig. 12.4). Mbp1 and Swi4 are also related in a C-terminal domain that is responsible for binding Swi6. All three of these proteins contain four ankyrin repeats within a central domain, which may indicate a common ancestry. Ankyrin repeats generally mediate protein–protein interactions, but their targets are unknown in these particular instances. Although MBF and SBF are thought normally to activate distinct sets of promoters, some crossregulation can occur. For example, in *Swi4⁻* mutants, *CLN1* and *CLN2* can be expressed under the influence of MBF, albeit less efficiently. However, *Mbp⁻ Swi4⁻* double mutants are inviable, due to a failure to synthesize Cln1 and Cln2.

Factors that resemble SBF and MBF are found in *Schizosaccharomyces pombe*. In terms of evolution, this fission yeast is as far removed from *S. cerevisiae* as either is from humans, as judged by sequence comparisons of a number of shared genes. Comparison of these two phylogenetically remote organisms can therefore provide insight into the most conserved and therefore fundamental aspects of cell cycle regulation. The cdc10 product of *S. pombe* is essential for DNA synthesis and appears to be the functional equivalent of Swi6. Cdc10 interacts with two DNA-binding factors, called Res1 and Res2, that are clearly homologous to Swi4 and Mbp1 (Fig. 12.4). Furthermore, the DNA sequences recognized by Res1 and Res2 resemble the SBF- and MBF-binding

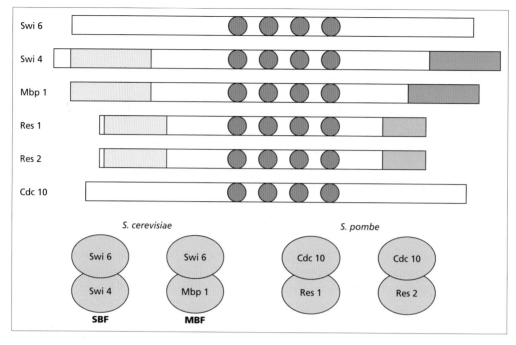

Fig. 12.4 Sequence similarities between SBF and MBF factors in *S. cerevisiae* and *S. pombe*. The pink N-terminal boxes indicate the DNA-binding domain that is conserved between Swi4, Mbp1, Res1 and Res2. The circles represent ankyrin repeats, of 33 amino acid residues each. The C-terminal coloured boxes in Swi4 and Mbp1 represent a conserved Swi6-binding domain, whereas those in Res1 and Res2 represent a conserved region that is required for interaction with cdc10. Beneath are indicated the complexes that form between these proteins.

sites in *S. cerevisiae*. The number of genes that become activated at the G1/S transition seems to be fewer in *S. pombe* than it is in *S. cerevisiae*, but they again code for a cyclin and enzymes involved in DNA synthesis.

The E2F family

Since the sequence and function of Swi4, Mbp1 and Swi6 have been conserved between budding and fission yeast, it seemed likely that homologues would also be found in higher eukaryotes that are no more remote from a phylogenetic point of view. Contrary to expectations, however, this extended family of proteins has only been detected in fungi. The closest functional equivalent of the Swi4/6 factors that has been identified in metazoa is the E2F family. Human E2F regulates transcription at the G1/S transition and binds to DNA sequences that closely resemble the binding sites of SBF and MBF in yeast. Furthermore, its target genes include several that encode cyclins and enzymes required for DNA synthesis. However, there is very little homology between the E2F proteins and Swi4 or Swi6 at the protein sequence level.

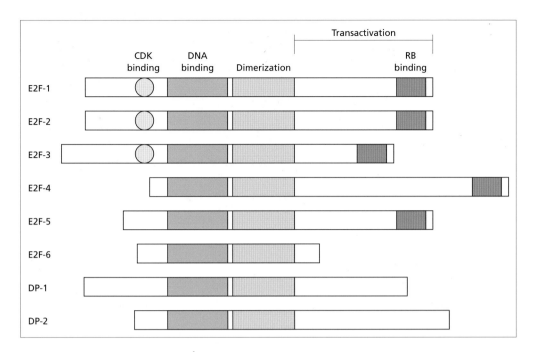

Fig. 12.5 Primary organization of the mammalian E2F and DP polypeptides. The conserved DNA-binding and dimerization domains are shaded pink. The RB-binding domain is marked red. A binding site for cdk2/cyclin A is present in E2F-1, -2 and -3, as indicated.

Six members of the E2F family have been discovered to date in mammals (Fig. 12.5). Efficient DNA binding by these proteins requires heterodimerization with a member of the DP family, of which there are two in mammals. The DPs are distantly related to the E2Fs, with a short stretch of homology in their DNA-binding and dimerization domains. *Drosophila* contain at least two E2F and one DP polypeptide. Recognition sites for E2F have been found in many genes that are involved in controlling proliferation. Examples include several genes that help to drive the cell cycle, such as cyclin E and the oncogene B-Myb. In addition, many of the enzymes required for DNA synthesis are encoded by genes with potential E2F sites, including thymidylate synthase and the replication origin-binding protein ORC1. In several of these cases, the E2F site has been shown to confer repression of the promoter during G0 and early G1 phase, followed by activation at the G1/S transition. This pattern of cell cycle control is mediated through protein–protein interactions between DNA-bound E2F and members of the RB family.

The RB family consists of three polypeptides in mammals; RB itself, p107 and p130. These nuclear phosphoproteins share several regions of homology, the most extensive of which is a large domain called the 'pocket' (Fig. 12.6). The pocket is a bipartite domain, composed of two well-conserved regions, A and B, that are highly intolerant of mutation, but which are separated by a

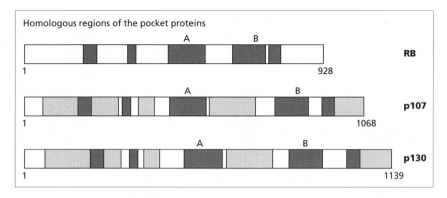

Fig. 12.6 Primary organization of the mammalian pocket proteins. Regions of homology between all three are shown as red boxes. p107 and p130 are more related to each other (~50% amino acid identity) than they are to RB (30–35% amino acid identity); regions shared by these two proteins that are not found in RB are grey. The A and B boxes of the pocket are indicated, as are the number of amino acid residues within each of these polypeptides.

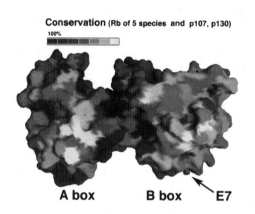

Fig. 12.7 Structure of the pocket region of human RB, as revealed by X-ray crystallography. The site of binding of the E7 oncoprotein of human papillomavirus is indicated. Reproduced with permission from *Nature* **391**. Copyright Macmillan Magazines Limited.

spacer region of variable length that will withstand extensive substitution. Both the A and B boxes contain a five-helix structural motif that is also found in the cyclins and is referred to as the 'cyclin fold' (Fig. 12.7). A similar cyclin fold arrangement is found in the basal pol II factor TFIIB. Crystallographic analysis suggests that the A box serves as a scaffold for the B box, which provides a surface that can bind to a range of viral and cellular proteins. One of the cellular targets for the pocket is the transcription activation domain of E2F. E2F-1, E2F-2 and E2F-3 bind almost exclusively to RB, whereas E2F-4 and E2F-5 can bind to all three pocket proteins. This interaction masks the E2F activation domain and blocks its ability to recruit TFIID and thereby stimulate transcription. E2F-6 lacks an activation domain and may serve as a competitive repressor of E2F-dependent transcription without any intervention from the pocket proteins.

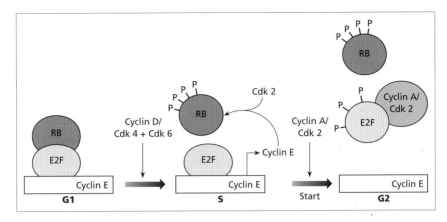

Fig. 12.8 Cell cycle regulation of an E2F-responsive gene, such as the *cyclin E* gene. During G0 and early G1 phases, the pocket proteins RB and p130 associate with the activation domain of promoter-bound E2F; this blocks its ability to stimulate transcription and recruits the histone deacetylase HDAC1, which may keep the promoter in a repressed chromatin state. External growth factors trigger the synthesis of D-type cyclins, which activate the cyclin-dependent kinases cdk4 and cdk6. As cells pass the restriction point, towards the end of G1 phase, cdk4 and cdk6 phosphorylate the pocket proteins at multiple sites, which inactivates and releases them, allowing the expression of E2F-responsive genes. The resultant synthesis of cyclin E activates the cyclin-dependent kinase cdk2, thereby providing a positive feedback loop that further phosphorylates the pocket proteins. Towards the end of S phase, cyclin A accumulates and replaces cyclin E as the partner for cdk2. The cyclin A/cdk2 pair binds to E2F-1, -2 and -3 through a conserved interaction domain and phosphorylates both components of the E2F/DP dimer; this inhibits DNA binding and silences expression of E2F-dependent genes.

Apart from its ability to bind and mask the activation domains of E2F proteins, RB can also counteract the effects of other stimulatory factors bound at nearby sites. In this way it can function as a dominant repressor of promoters to which it is recruited. One way in which it appears to do this is by recruiting the histone deacetylase HDAC1. As described in Chapter 8, this is likely to remove acetyl groups from neighbouring histones and thereby convert promoters into closed chromatin states. For this reason, E2F recognition motifs frequently serve as inhibitory elements. However, the pocket proteins only function in this way during G0 and early G1 phases. As cells reach the restriction point, the mammalian equivalent of Start, the pocket proteins undergo extensive phosphorylation at multiple sites by the cyclin D-dependent kinases cdk4 and cdk6. This inactivates them and causes them to dissociate from E2F, thereby allowing the expression of E2F-responsive genes (Fig. 12.8). One of the genes that was repressed by the pocket proteins encodes cyclin E; this binds and activates the cyclin-dependent kinase cdk2, which further phosphorylates RB and its relatives. This provides a positive feedback loop, which ensures that the initial phosphorylation of pocket proteins by cdk4 and cdk6 is reinforced rapidly by cdk2 to give a concerted transition. The promoter of the

E2F-1 gene also contains an E2F recognition sequence and therefore becomes activated when the pocket proteins are switched off, thereby providing further positive feedback. The net effect is that passage through the restriction point is accompanied by a substantial increase in the expression of a battery of genes that contribute to proliferation. The key role of E2F in this process is illustrated by the fact that its overexpression is sufficient to drive progression from G0 into S phase in certain cell types.

Exit from S phase appears to require the inactivation of E2F, which may be achieved through two distinct mechanisms. One involves cyclin A, which is synthesized in S phase following the wave of cyclin E production (Fig. 12.1). Cyclin A binds and activates cdk2, and this complex interacts with a conserved motif in E2F-1, E2F-2 and E2F-3. Once they have associated, cyclin A/cdk2 phosphorylates both the E2F and DP components of the heterodimer and inhibits their ability to bind to DNA. Expression of an E2F-1 mutant without the cdk-interaction site or a DP-1 mutant lacking its major phosphoacceptor sites stabilizes the E2F complex and causes cells to arrest in S phase. The second mechanism which may contribute to the inactivation of E2F in S phase involves its ubiquitin-dependent proteolytic degradation. Sequences in the activation domain of E2F-1 and E2F-4 are targetted by the ubiquitin-proteasome machinery and destabilize these proteins. These sequences are masked by the pocket proteins during G0 and early G1 phases, giving the E2F-repressor complexes much greater stability.

Retinoblastoma protein

Retinoblastoma (RB) is extremely abundant for a transcription factor and it has been estimated there may be 100 RB molecules for every E2F molecule in some mammalian cell types. A wide variety of proteins besides E2F have been shown to bind to RB, which provides an explanation for its high concentration. Table 12.1 lists some of the known RB-binding proteins, many of which are transcription factors. Two interesting examples are provided by the pol I-specific factor UBF and the pol III-specific factor TFIIIB, both of which become repressed following their association with RB. Like TBP, RB has the rare ability to regulate transcription by all three nuclear RNA polymerases (Fig. 12.9). Because UBF and TFIIIB are believed to contribute to all transcription by pols I and III, respectively, targetting these factors provides RB with the opportunity of repressing all class I and class III genes. In the latter case this is a huge number of templates, because there are 400 5S rRNA genes in a haploid human genome, 1300 tRNA genes and half a million Alu genes. Targetted disruption of the *Rb* gene in mice results in a five-fold increase in the rate of synthesis of tRNA and 5S rRNA, which must make a tremendous difference to the overall level and balance of nuclear activity.

RB is believed to be a principal executor of the restriction point, the mammalian equivalent of Start. Overexpression of RB can cause cells to arrest in G1

Table 12.1 Proteins that bind to RB.

RB-binding protein	Function
Large T antigen	SV40 oncoprotein
E1A	Adenovirus oncoprotein
E7	HPV oncoprotein
UBF	Transcription factor—pol I
E2F	Transcription factor—pol II
ATF2	Transcription factor—pol II
PU.1	Transcription factor—pol II
C/EBP	Transcription factor—pol II
MyoD	Transcription factor—pol II
Myogenin	Transcription factor—pol II
TAF$_{II}$250	Transcription factor—pol II
TFIIIB	Transcription factor—pol III
Abl	Tyrosine kinase
D-type cyclins	Regulators of cdk4 and cdk6 kinases
BRG family	Remodel chromatin
HDAC1 and 2	Remodel chromatin

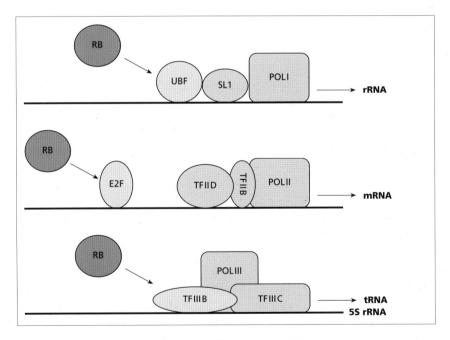

Fig. 12.9 RB can regulate transcription by all three nuclear RNA polymerases. It represses pol I transcription by binding to UBF. It can influence pol II transcription both positively and negatively, depending on circumstances, through interactions with a variety of regulatory factors, such as E2F. It inhibits pol III transcription by binding and repressing TFIIIB.

phase. The decision as to whether or not a mammalian cell can progress through the restriction point and commit itself to another round of replication and division is made in response to external stimuli such as growth factors. Mitogenic growth factors stimulate the synthesis of D-type cyclins, which bind to the kinases cdk4 and cdk6 and stimulate these to phosphorylate and inactivate RB. Once it is hyperphosphorylated, RB dissociates from many of its targets, including E2F and TFIIIB. Release of E2F from repression allows the synthesis of proteins that promote cell cycle progression, such as cyclin E. Release of TFIIIB and UBF from repression increases the production of tRNA and rRNA, which may be necessary for cells to achieve a sufficient rate of protein synthesis to sustain active growth. By controlling these disparate processes, RB may serve as a master switch which allows a potent, pleiotropic and coordinated cellular response to the external presence of mitogenic signals. Key messengers in this process are the D-type cyclins, which are synthesized continuously as long as growth factor stimulation persists. They are degraded very rapidly if growth factors are withdrawn, leading to the reactivation of RB and G1 phase arrest. The role of the D-type cyclins is clearly analogous to that of Clns1–3 in yeast. In both cases their overexpression can contract the duration of G1 phase, reduce cell size and diminish cellular dependence on growth-promoting signals.

In addition to its role as a suppressor of growth and proliferation, RB can also function to promote cellular differentiation. In some respects these two activities may be regarded as opposite sides of the same coin, as differentiating cells generally exit from the cycle and stop growing. Several cell types fail to differentiate properly in $Rb^{-/-}$ mice, including lens, neuronal and erythroid precursors. There is also evidence that RB is involved in the formation of muscle, bone and adipose tissue. $Rb^{-/-}$ myotubes, unlike their wild-type counterparts, fail to withdraw from the cell cycle. The underphosphorylated form of RB has been shown to bind to the bHLH domains of the myogenic factors MyoD and myogenin. This interaction may enhance the ability of these bHLH proteins to induce myogenesis and activate muscle-specific markers, such as the muscle creatine kinase gene. Similarly, RB can promote adipocyte differentiation in culture, an effect that may involve its interaction with members of the CCAAT/enhancer binding protein (C/EBP) family of bZIP proteins. RB associates with C/EBP proteins during adipogenesis and enhances their DNA-binding and transcriptional activities. Several other examples have also been reported in which RB potentiates transcriptional activation. Thus, in addition to regulating all three nuclear RNA polymerases, this remarkable and pluripotent protein can serve to either increase or decrease transcription, depending on target and context.

Because of its important roles as an inhibitor of growth and proliferation and a promoter of cellular differentiation, RB can function as a tumour suppressor. Pathological examination of human tumours reveals that proliferation and differentiation are both frequently deregulated and generally show an inverse correlation. Overexpression of exogenous *Rb* in cancer cells can inhibit growth, proliferation and tumorigenicity. Following the targetted disruption of one *Rb* allele, >95% of heterozygous mice die of cancer within 300–400 days of birth. RB is frequently found to be mutated in human tumours, thereby freeing cancer cells from its restraining influence and allowing deregulated growth and proliferation. Individuals who inherit a non-functional *Rb* allele have an approximately 90% chance of developing retinoblastoma at an early age, following somatic mutation of the remaining allele. Patients who survive hereditary retinoblastoma display a strong predisposition to osteosarcomas and soft tissue sarcomas later in life; this is again associated with loss of the second *Rb* allele. Many other types of human tumour display somatic mutations in *Rb*, including breast, prostate, bladder and lung cancers. In such cases, the patient inherits two wild-type copies of the *Rb* gene, but mutations arise in both alleles during tumour development. The most striking examples of this are the small cell lung carcinomas, where *Rb* changes are found in nearly all cases. Other tumour types display a lower frequency of *Rb* mutation; for example, *Rb* was found to be altered or absent in a third of bladder carcinomas surveyed.

Some human cervical cancers display inactivating mutations in *Rb*. However, the majority contain wild-type *Rb* but are also infected by human papillomavirus (HPV), which is the principle aetiological agent. HPVs encode an oncoprotein called E7, which can transform cells and can bind to RB. Some HPV types, such as HPV-16 and HPV-18, are associated with potentially pre-cancerous genital tract lesions and a large percentage of anogenital cancers, whereas others (e.g. HPV-6 and HPV-11) are associated with benign proliferative tumours with a low risk of malignant progression (e.g. condyloma acuminata). E7 proteins from the high-risk viruses HPV-16 and HPV-18 have higher binding affinity for RB than E7 from the lower risk types HPV-6 and HPV-11. Single-residue substitutions in HPV-6 E7 that cause a substantial increase in affinity for RB also produce a concomitant gain in transforming ability. It is therefore highly likely that the ability of E7 to bind RB contributes substantially to the oncogenic capacity of HPVs. RB function may therefore be lost in most if not all cervical cancers; this occurs by gene mutation in the minority of HPV-negative cases and by complex formation with E7 in the remaining cases.

Several other DNA tumour viruses encode transforming proteins that can bind RB and neutralize its function. Examples include the E1A protein of adenovirus and the large T antigen of simian virus 40. Mutations in these oncoproteins that prevent RB binding also abolish their capacity to transform. By associating with RB, these viral proteins can interfere with its normal cellular functions and thereby mimic the effects of the *Rb* mutations that occur in many tumours. The same viruses have also evolved mechanisms to inactivate the key tumour suppressor p53 (Box 12.1).

As already described, inactivation of RB through phosphorylation by the cyclin D- and E-dependent kinases constitutes a normal control mechanism

Box 12.1 DNA tumour viruses, RB and p53

Adenovirus, simian virus 40 (SV40), human papillomavirus (HPV) and
human herpesvirus 8 (HHV8) have each evolved products that allow
them to neutralize two of the cell's key defences, RB and p53. The
pocket domains of RB, p107 and p130 are bound by the E1A product of
adenovirus, the E7 product of HPV and the large T antigen of SV40; this
prevents the pocket proteins from interacting with many of their key
cellular targets, such as E2F and TFIIIB. The same end is achieved
through hyperphosphorylation in the case of HHV8, a causative agent
of Kaposi's sarcoma and body cavity lymphomas (it is also called
Kaposi's **s**arcoma-associated **h**erpes**v**irus or KSHV). HHV8 encodes its
own cyclin, which binds and activates the kinase cdk6, causing it to
phosphorylate and switch off the pocket proteins. The viral cyclin is
especially potent, because it is resistant to inhibitors such as p16 and
p21 that regulate the cellular cyclin/cdk complexes. HHV8 also
produces a latency associated nuclear antigen that binds and represses
p53. Similarly, p53 is bound and neutralized by the E1B product of
adenovirus, which contacts its transcriptional activation domain. SV40
large T antigen and the E6 oncoprotein of HPV interact with the DNA-
binding domain of p53; in the latter case, this targets p53 for rapid
proteolytic degradation. The recurring theme of inhibition of both RB
and p53 highlights the importance of these tumour suppressors in
protecting cells. The simplest explanation for this dual requirement is
that the viral oncoproteins which inactivate RB drive growth and
proliferation, whereas the oncoproteins that target p53 neutralize the
defence mechanisms which would otherwise respond to abnormal
replication (viral or cellular) and trigger apoptosis in the infected cell.
For example, expression of E7 alone in transgenic mice causes
abnormal proliferation and differentiation, but the targetted tissues
degenerate due to apoptosis, rather than developing tumours. By
contrast, expression of E7 in a $p53^{-/-}$ mouse, or coexpression of E7 with
E6, provokes tumour formation rather than programmed cell death.
Efficient immortalization of primary genital epithelial cells requires
both E6 and E7. The same rationale can explain why many cancer cell
types that are not infected with DNA tumour viruses carry mutations in
the genes for both RB and p53.

that is used to regulate progress through the cell cycle. Unfortunately, this
process is deregulated in many human malignancies, thereby switching off
wild-type RB in situations in which it would otherwise be functional. The gene
encoding cyclin D1 has been found to be amplified in at least 15% of primary

breast cancers and an even higher proportion of squamous cell carcinomas of the lung, neck, head and oesophagous. This gene can also be overexpressed due to other types of aberration, including proviral insertions and chromosomal rearrangements, and it is classified as a proto-oncogene. Cyclin D2 can also be activated by analogous abnormalities and the gene encoding cdk4 is amplified in many glioblastomas. In addition to these many cases in which the D-type cyclins and their associated kinase are upregulated directly, many other cancers lose the function of p16, which is an important repressor of cyclin D-dependent kinases. For example, the gene for p16 is deleted somatically in many oesophageal, bladder, lung and pancreatic carcinomas and is sometimes mutated in familial melanomas. In a number of malignancies, this gene has been found to have been silenced inappropriately due to the hypermethylation of its promoter. Furthermore, some melanomas have mutations in cdk4 that prevent it from interacting with p16. Thus, a diversity of mechanisms can result in the abnormal elevation of cyclin D-dependent kinase activity in a broad spectrum of cancers. This has the effect of switching off RB.

It is therefore certain that RB function is lost in a high proportion of tumours. Indeed, it has been suggested that the restriction point control pathway involving RB may become deregulated in all human malignancies. This can be achieved in a variety of different ways—gene mutation, association with viral oncoproteins or hyperphosphorylation (Fig. 12.10). A good illustration of the importance of inactivating RB during tumour progression was provided by a survey of small cell lung carcinomas: of 55 cases tested, 48 were found to lack normal RB but contain wild-type p16, and six of the remaining seven lacked functional p16. This study also illustrates the point that it is rare for a tumour to contain defects in more than one component of the pathway; if p16 is lost, RB is switched off through hyperphosphorylation and there is no pressure for it to suffer mutation. The end point is similar in each case—a deregulation of restriction point control that may be obligatory for tumorigenesis.

Repression of transcription at mitosis

In higher eukaryotes, mitosis is associated with an abrupt inhibition of transcription. This was first noted in autoradiographic studies which found that the incorporation of radioactive precursors into nuclear RNA decreases in prophase and resumes in late telophase at the end of mitosis. All three nuclear RNA polymerases are subject to mitotic inhibition and a surprising variety of molecular mechanisms are believed to contribute to this effect.

One of the most striking features of an M phase cell is the dramatic condensation of chromosomes. A priori, this seemed to provide a very clear mechanism for masking and silencing genes. Several studies have found that the occupancy of specific promoter or enhancer regions is altered in mitotic chromosomes. For example, in vivo footprinting has revealed that several

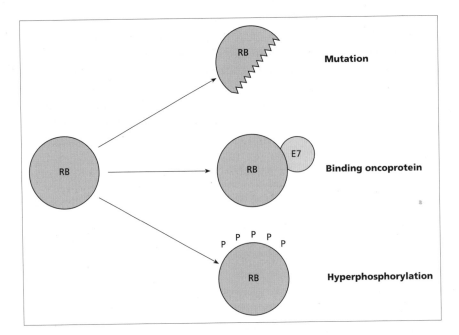

Fig. 12.10 Three different types of mechanism can account for the loss of RB function in transformed and tumour cells. The *Rb* gene is mutated in many types of tumour, such as retinoblastomas. Wild-type RB can be neutralized by the binding of a viral oncoprotein, such as the E7 protein of HPVs in cervical cancers. Wild-type RB can also be neutralized through hyperphosphorylation by the cyclin D-dependent kinases cdk4 and cdk6; this can result from the overexpression of cyclin D, as is found in many breast cancers, or the mutation of p16, as occurs in some cases of familial melanoma.

factors are released from the human *hsp70* promoter during mitosis, including C/EBP, Sp1 and heat shock factor 1 (HSF1). Immunocytochemical analyses have shown that pol II and many regulatory factors are dispersed from bulk chromatin during M phase. In some cases, such as Sp1 and Oct-1, transcription factor release correlates with a decrease in its DNA-binding capability. However, the DNA-binding activity of C/EBP and HSF1 appears not to change during mitosis, which suggests that these factors may be actively displaced by the process of chromatin compaction. It is unclear how this is achieved, but there is substantial biochemical and immunocytochemical evidence for a net deacetylation of the core histones in mitotic chromosomes. Histones H1 and H3 also become hyperphosphorylated at mitosis, but the significance of this is unclear. Human SWI/SNF complexes are hyperphosphorylated too during this phase of the cell cycle; their remodelling functions are inactivated and they are excluded from chromatin.

As already mentioned, the DNA-binding activity of Sp1 and Oct-1 is diminished during M phase. It is possible that this contributes to the displacement of neighbouring transcription factors that are not regulated directly, either by a loss of cooperative interactions or by alterations in the local configuration of chromatin that disfavour factor occupancy. For both Sp1 and Oct-1, the mitotic repression of DNA binding is due to hyperphosphorylation. Oct-1 is phosphorylated at several sites, but Ser-385 is only labelled during M phase. This residue is located within the N-terminal arm of the POU homeodomain and the presence of a phosphate here is likely to interfere with DNA recognition. This phosphoacceptor site is conserved in all POU domain proteins and may therefore provide a common mechanism for regulating this family. Indeed, the DNA-binding capability of Pit-1 has been shown to be inhibited at mitosis due to phosphorylation at the corresponding site.

Inactivating essential components of the basal transcription machinery should provide a highly efficient mechanism for achieving global repression. Indeed, TATA-binding protein (TBP) becomes hyperphosphorylated during mitosis, as do TAF components of SL1, TFIID and TFIIIB. In each case this is responsible for a substantial loss of function. For pols I and III, this may be sufficient to account for the downregulation of expression, but the effect is more selective for TFIID, because the response to activators is inhibited more dramatically than basal pol II transcription. Although multiple subunits of TFIID are hyperphosphorylated at mitosis, the complex appears to remain intact.

Immunocytochemistry and subcellular fractionation indicate that the majority of TFIID is displaced into the cytoplasm from the disassembling prophase nucleus around the time of nuclear envelope breakdown. Nevertheless, 10–20% of TFIID remains associated with the condensed mitotic chromosomes. SL1, UBF, AP-2 and SRF have also been shown to be retained on chromosomes during M phase. The selective retention of specific transcription factors might serve to mark particular genes for rapid reactivation at the start of interphase; otherwise the entire pattern of gene expression would need to be reprogrammed after each mitosis. The idea that previously active genes can be marked during M phase is supported by the fact that the human *hsp70* promoter remains hypersensitive to DNase I, despite the displacement of several regulatory factors. Indeed, the TATA box and start site region of this gene stays unwound during mitosis, as revealed by footprinting with potassium permanganate. The same is true of other promoters that were active prior to M phase, but is not found with genes that were already silent in the previous interphase.

During interphase, cells can respond to various stresses with a range of protective transcriptional programmes. The general abrogation of transcription that occurs during mitosis deprives cells of this response mechanism and makes them especially vulnerable to environmental insults. A clear illustration of this is provided by the heat shock response, in which the rapid induction of several genes, such as *hsp70*, is critical in allowing cells to survive a range of stress conditions. When an asynchronous population of HeLa cells was subjected to an acute heat shock (1 hour at 43°C), 72% were found to survive; this number fell to 5% when the population was first synchronized in M phase. It is extremely fortunate that mitosis is the shortest of the cell cycle phases!

Further reading

Reviews

Breeden, L. (1996) Start-specific transcription in yeast. *Curr. Top. Microbiol. Immunol.* **208**: 95–127.

Cross, F. R. (1995) Starting the cell cycle: what's the point? *Curr. Opin. Cell Biol.* **7**: 790–797.

Dynlacht, B. D. (1997) Regulation of transcription by proteins that control the cell cycle. *Nature* **389**: 149–152.

Dyson, N. (1998) The regulation of E2F by pRB-family proteins. *Genes Dev.* **12**: 2245–2262.

Gottesfeld, J. M. & Forbes, D. J. (1997) Mitotic repression of the transcriptional machinery. *Trends Biochem. Sci.* **22**: 197–202.

Helin, K. (1998) Regulation of cell proliferation by the E2F transcription factors. *Curr. Opin. Genet. Dev.* **8**: 28–35.

Herwig, S. & Strauss, M. (1997) The retinoblastoma protein: a master regulator of cell cycle, differentiation and apoptosis. *Eur. J. Biochem.* **246**: 581–601.

Koch, C. & Nasmyth, K. (1994) Cell cycle regulated transcription in yeast. *Curr. Opin. Cell Biol.* **6**: 451–459.

Mulligan, G. & Jacks, T. (1998) The retinoblastoma gene family: cousins with overlapping interests. *Trends Genet.* **14**: 223–229.

Strauss, M., Lukas, J. & Bartek, J. (1995) Unrestricted cell cycling and cancer. *Nature Med.* **12**: 1245–1246.

Taya, Y. (1997) RB kinases and RB-binding proteins: new points of view. *Trends Biochem. Sci.* **22**: 14–17.

Vousden, K. H. (1995) Regulation of the cell cycle by viral oncoproteins. *Semin. Cancer Biol.* **6**: 109–116.

White, R. J. (1997) Regulation of RNA polymerases I and III by the retinoblastoma protein: a mechanism for growth control? *Trends Biochem. Sci.* **22**: 77–80.

Selected papers

Cell cycle regulation in Saccharomyces

Koch, C., Schleiffer, A., Ammerer, G. & Nasmyth, K. (1996) Switching transcription on and off during the yeast cell cycle: Cln/Cdc28 kinases activate bound transcription factor SBF (Swi4/Swi6) at Start, whereas Clb/Cdc28 kinases displace it from the promoter in G2. *Genes Dev.* **10**: 129–141.

Polymenis, M. & Schmidt, E. V. (1997) Coupling of cell division to cell growth by translational control of the G1 cyclin *CLN3* in yeast. *Genes Dev.* **11**: 2522–2531.

E2F

Bandara, L. R. & La Thangue, N. B. (1991) Adenovirus E1a prevents the retinoblastoma gene product from complexing with a cellular transcription factor. *Nature* **351**: 494–497.

Chellappan, S. P., Hiebert, S., Mudryj, M., Horowitz, J. M. & Nevins, J. R. (1991) The E2F transcription factor is a cellular target for the Rb protein. *Cell* **65**: 1053–1061.

DeGregori, J., Kowalik, T. & Nevins, J. R. (1995) Cellular targets for activation by the E2F1 transcription factor include DNA synthesis- and G1/S-regulatory genes. *Mol. Cell. Biol.* **15**: 4215–4224.

Krek, W., Xu, G. & Livingston, D. M. (1995) Cyclin A-kinase regulation of E2F-1 DNA binding function underlies suppression of an S phase checkpoint. *Cell* **83**: 1149–1158.

Yamasaki, L., Jacks, T., Bronson, R., Goillot, E., Harlow, E. & Dyson, N. J. (1996) Tumour induction and tissue atrophy in mice lacking E2F-1. *Cell* **85**: 537–548.

Zwicker, J., Liu, N., Engeland, K., Lucibello, F. C. & Muller, R. (1996) Cell cycle regulation of E2F site occupation *in vivo*. *Science* **271**: 1595–1597.

Retinoblastoma protein

Cavanaugh, A. H., Hempel, W. M., Taylor, L. J., Rogalsky, V., Todorov, G. & Rothblum, L. I. (1995) Activity of RNA polymerase I transcription factor UBF blocked by Rb gene product. *Nature* **374**: 177–180.

Gu, W., Schneider, J. W., Condorelli, G., Kaushal, S., Mahdavi, V. & Nadal-Ginard, B. (1993) Interaction of myogenic factors and the retinoblastoma protein mediates muscle cell commitment and differentiation. *Cell* **72**: 309–324.

Luo, R. X., Postigo, A. A. & Dean, D. C. (1998) Rb interacts with histone deacetylase to repress transcription. *Cell* **92**: 463–473.

Sellers, W. R., Novitch, B. G., Miyake, S. *et al.* (1998) Stable binding to E2F is not required for the retinoblastoma protein to activate transcription, promote differentiation, and suppress tumour cell growth. *Genes Dev.* **12**: 95–106.

Weintraub, S. J., Prater, C. A. & Dean, D. C. (1992) Retinoblastoma protein switches the E2F site from positive to negative element. *Nature* **358**: 259–261.

White, R. J., Trouche, D., Martin, K., Jackson, S. P. & Kouzarides, T. (1996) Repression of RNA polymerase III transcription by the retinoblastoma protein. *Nature* **382**: 88–90.

Mitotic repression of transcription

Gottesfeld, J. M., Wolf, V. J., Dang, T., Forbes, D. J. & Hartl, P. (1994) Mitotic repression of RNA polymerase III transcription *in vitro* mediated by phosphorylation of a TFIIIB component. *Science* **263**: 81–84.

Martinez-Balbas, M. A., Dey, A., Rabindran, S. K., Ozato, K. & Wu, C. (1995) Displacement of sequence-specific transcription factors from mitotic chromatin. *Cell* **83**: 29–38.

Michelotti, E. F., Sanford, S. & Levens, D. (1997) Marking of active genes on mitotic chromosomes. *Nature* **388**: 895–899.

Segil, N., Boseman Roberts, S. & Heintz, N. (1991) Mitotic phosphorylation of the Oct-1 homeodomain and regulation of Oct-1 DNA binding activity. *Science* **254**: 1814–1816.

Segil, N., Guermah, M., Hoffmann, A., Roeder, R. G. & Heintz, N. (1996) Mitotic regulation of TFIID: inhibition of activator-dependent transcription and changes in subcellular localization. *Genes Dev.* **10**: 2389–2400.

White, R. J., Gottlieb, T. M., Downes, C. S. & Jackson, S. P. (1995) Mitotic regulation of a TATA-binding-protein-containing complex. *Mol. Cell. Biol.* **15**: 1983–1992.

Chapter 13: Interactions Between Transcription and Other Nuclear Processes

Inevitably, transcription factors and transcription itself can have an important influence upon other activities occurring within the nucleus. Major examples that will be described in this chapter include the processing of transcripts, the replication of chromosomes and the repair of certain types of DNA damage.

Transcription and RNA processing

To some extent, the post-transcriptional fate of RNAs is determined by which polymerase is responsible for making them. As it is synthesized, a pol II transcript generally undergoes a number of modifications that help determine its stability, translatability and often, by alternative splicing, the nature of the protein it encodes. A primary transcript that will give rise to mRNA when made by pol II, will not be processed in the same way if it is made by pols I or III or a bacteriophage RNA polymerase. This is due to a unique feature of pol II, the C-terminal domain (CTD)—a conserved repeat of the heptad YSPTSPS that is found at the C-terminus of its largest subunit (see Chapter 2). The CTD binds to many of the proteins that are involved in capping, splicing and cleaving the transcript at a poly(A) site. None of these modifications occurs efficiently if the CTD is truncated. In essence, the CTD serves to target the mRNA maturation apparatus specifically to the products of pol II.

The first step in the processing pathway is for the 5′ end of the pre-mRNA to be modified by capping. This occurs shortly after initiation, before the transcript has reached 40 nucleotides in length. It involves cleavage of the terminal phosphate, addition of GMP by a 5′–5′ triphosphate bridge to give GpppN, and then methylation of the G at its N7 position to generate an m^7G cap. In both yeast and man, the capping enzymes are found to associate specifically with the CTD when it has been phosphorylated. As was explained in Chapter 4, CTD phosphorylation occurs at the time of transcript initiation and is maintained during elongation. The capping complex does not associate with pol II before it starts to transcribe. Binding to the CTD shortly after initiation may serve to direct the capping enzymes to the 5′ ends of nascent transcripts. Furthermore, the phosphorylated CTD activates the capping enzyme by raising its affinity for GTP. mRNA capping is decreased by five-fold in human cells in which pol II has had its CTD truncated to five heptad repeats instead of the normal 52. The fact that transcripts of pols I and III are not capped can be attributed to the absence of a CTD in these polymerases.

Splicing is a transesterification reaction that removes intronic sequences from RNA and ligates together the regions encoded by exons. The process is carried out by a multiprotein complex called the spliceosome that includes snRNAs. Cytological studies have shown that spliceosomes begin to interact with 5′ splice sites before 3′ splice sites have even been synthesized, and that introns are often removed while elongating transcripts are still attached to the DNA; in other words, it occurs whilst a gene is still being transcribed. As is the case for capping, pol II with a truncated CTD is deficient for splicing. Overexpression of the CTD in HeLa cells interferes with splicing *in vivo* and immunodepletion of pol II from cell-free extracts inhibits splicing *in vitro*. These effects can be explained by the observation that the hyperphosphorylated CTD interacts with splicing snRNPs and other splicing factors. Although this is likely to help direct the spliceosome to nascent RNA, the transcript itself must also contribute significantly to the binding specificity of the splicing apparatus. Splicing factors were found to associate selectively with transfected expression constructs, only binding to those that contain introns, even though intronless templates would still attract pol II. Although nearly every mRNA is capped and polyadenylated, not all are spliced; for example, in *S. cerevisiae* only 235 of the ~6000 genes are spliced, a much smaller proportion than occurs in mammals. Furthermore, some pol III transcripts undergo splicing, such as certain tRNAs. The number of molecules of snRNPs and many other splicing factors far exceeds the number of CTDs in a cell, consistent with the existence of CTD-independent pathways for intron removal.

As for splicing, sequences in the nascent RNA undoubtedly play an important role in directing the polyadenylation factors to the appropriate sites. However, the pol II transcription apparatus helps to direct these factors to their target sequences (AAUAAA and GU-rich tracts, see Chapter 4) and RNAs synthesized by pols I and III are not generally polyadenylated. Transcripts synthesized *in vivo* by pol II with a truncated CTD are not cleaved efficiently at the poly(A) site. Furthermore, immunofluorescence analysis has shown that the cleavage-stimulation factor CstF colocalizes with hyperphosphorylated pol II in living *Drosophila* cells. The CTD facilitates the 3′ RNA cleavage reaction *in vitro*, even in the absence of ongoing transcription. This capacity can be explained by the fact that the CTD binds to CstF and the polyadenylation specificity factor CPSF. It has been suggested that the CTD may function as a cofactor or allosteric activator for the 3′ end processing reaction. In addition, CstF and CPSF also bind to some of the TATA-binding protein associated factor (TAF) subunits of TFIID, a somewhat unexpected finding because TFIID is left behind at the promoter once transcription commences. It appears that CPSF is recruited to class II genes through interaction with TFIID and is then transferred to the CTD during the initiation process (Fig. 13.1). Although these interactions with the transcription machinery help target the polyadenylation apparatus, they are not able to override the processing signals that are contained within the sequence of the RNA. Thus, if a construct is engineered so

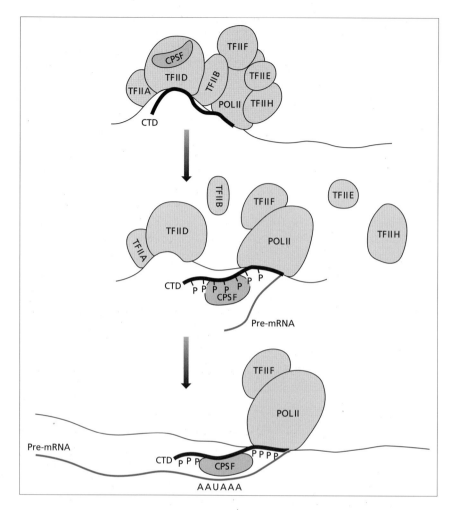

Fig. 13.1 Model depicting the fate of cleavage-polyadenylation specificity factor (CPSF) during transcript initiation, elongation and 3'-mRNA processing. CPSF is recruited to the promoter through interactions with several of the TAF subunits of TFIID. When transcription begins, it dissociates from TFIID and binds instead to the CTD of pol II. As the polyadenylation site is reached, CPSF may dissociate from pol II and direct the processing of the 3' end of the mRNA.

that pol II transcribes an rRNA gene, the product is processed like a normal rRNA and not as an mRNA.

Because pol II helps control the mRNA processing reaction, it is possible that the communication works in both directions and that the processing factors can influence the decision between continued transcript elongation or termination. Signals to pol II from the polyadenylation apparatus may ensure that the transcript terminates only after the end of the gene has been reached; this might be achieved through conformational changes. More generally, the processing machinery might influence transcription in additional ways. A study using transgenic mice found that transcription from several natural promoters is stimulated by the presence of downstream introns. Perhaps the presence of an intron helps attract pol II via its interaction with the splicing factors. Signals from the processing machinery might be propagated through the CTD to coordinate transcript elongation with its processing, thereby providing a 'quality control' mechanism for mRNA surveillance. Such coupling might signal transcription to pause or terminate in response to delays or errors in the assembly or function of spliceosomes.

It seems clear from these data that a pol II transcription unit is closely associated with the capping, splicing and polyadenylation apparatus to form an mRNA synthesis factory. This may ensure the effective coordination of these activities *in vivo*. The CTD may act as a scaffold that concentrates the processing apparatus and ensures that it is positioned appropriately with respect to the transcript. There is also some indication that it may actively contribute to the 3' cleavage reaction, perhaps as an allosteric activator. Cytological examinations have shown that processing factors are highly concentrated at the sites of active pol II transcription, which are themselves distributed throughout the nucleoplasm. To some extent, then, nuclear organization is determined by gene transcription.

Transcription and DNA replication

There is considerable evidence that the transcription machinery can also influence the process of DNA replication. A clear illustration of this is provided by the fact that actively transcribed genes are replicated earlier in S phase than silent genes. This can even be true of the two copies of a particular gene within the same cell, if one copy is expressed and the other is not. A striking example of this phenomenon is provided by the 5S rRNA genes in *Xenopus*. As was described in Chapters 8 and 9, the oocyte type 5S rRNA genes are silenced in somatic cells, despite having functional promoters and displaying only six nucleotide differences from the actively transcribed somatic 5S rRNA genes. Under these circumstances, the repressed oocytic 5S rRNA genes are replicated late whereas the somatic genes are replicated early. However, it was found that in a kidney cell line a block of the oocyte genes has translocated to the nucleolar organizer region, a site of active transcription; not only are these genes expressed but they are also replicated early.

DNA replication initiates at discrete sites called origins that are recognized by the replication apparatus. The origin core is generally flanked by binding sites for transcription factors, which can stimulate replication by as much as a thousand-fold under certain experimental conditions. For example, many replication origins in *S. cerevisiae* are flanked by a binding site for ABF1, a transcription factor that is essential for viability. When the ABF1 site at a particular origin was replaced by the recognition sequence for the galactose-inducible factor Gal4, replication enhancement at that origin became galactose-dependent.

In some cases, such as the SV40 and polyoma viruses, the same binding sites can serve as enhancers of both transcription and DNA replication. Activation domains are sometimes required for a transcription factor to increase replication, but this is not invariably the case. Factors that can activate transcription very efficiently will not always stimulate replication from a particular origin. For example, the acidic activation domain of VP16 has no effect on the SV40 origin and only a slight effect on the origin of polyoma virus. By contrast, AP1 has a potent effect on both of these origins, despite being a much weaker transcriptional regulator than VP16. The ability of a given factor to activate DNA replication can also be dependent on cell type, which may reflect an involvement of cofactors.

Several different mechanisms, which are not mutually exclusive, can be used by transcription factors to regulate DNA replication. In adenovirus 2, the origin lies adjacent to binding sites for NF1 and Oct-1, both of which can stimulate replication independently of their activation domains. NF1 functions through a recruitment mechanism, binding to the adenoviral DNA polymerase and facilitating its interaction with the origin. By contrast, Oct-1 appears to function by promoting unwinding of the DNA duplex at the origin. In the SV40 genome, the replication origin lies close to a cluster of Sp1-binding sites, which form part of the viral early promoter (see Fig. 5.2). When bound to these sites, Sp1 is thought to promote unwinding at the origin. However, the protein responsible for initiating DNA unwinding is the SV40 product large T antigen. As well as being a replication factor, large T antigen is a transcription factor in its own right, regulating the expression of a range of viral and cellular genes. Another of its roles, as was described in Chapter 12, is to drive cells into S phase by binding and inactivating RB and p53, both of which can arrest cells in G1 phase. By promoting S phase entry, T antigen activates the cellular machinery for DNA synthesis, thereby allowing the coincident replication of the SV40 genome.

One of the principle mechanisms that allow transcription factors to stimulate replication is likely to involve their ability to influence chromatin structure. Chromatin has been shown to have a strong effect on replication origins in yeast, fruitfly and mammalian chromosomes. For example, heterochromatin can suppress the utilization of origins in both *Saccharomyces* and *Drosophila*. Incorporating part of a yeast origin into a nucleosome has also been shown to inhibit its use. The obvious explanation for these phenomena is that chromatin can impede the access of the DNA replication machinery. In many cases, transcription factors may help to maintain a more open chromatin structure. For example, SV40 replication can be inhibited if its origin is incorporated into nucleosomes, but this repression can be relieved by the presence of NF1 bound to an adjacent engineered site; this reflects the ability of NF1 to compete with histones, as NF1 does not increase replication of the same SV40 construct in the absence of nucleosomes. Similarly, AP-1 stimulates polyoma virus replication, but only under conditions where the DNA is assembled into a repressive chromatin structure. The E1 protein of human papillomavirus binds to the hSNF5 subunit of the SWI/SNF complex; this may allow it to recruit SWI/SNF to the viral replication origin and use its remodelling activity to facilitate chromatin opening.

Both the chromatin structure and transcriptional activity of the human β-globin cluster is dictated by its locus control region (LCR), as was described in Chapter 8; in addition, this LCR has been found to determine the timing of replication of the β-globin genes. Most of the LCR is deleted in patients with Hispanic thalassaemia. Not only does this prevent the expression of β-globin in erythroid cells, but it also shifts the time at which the locus is replicated from early to late in S phase. This is the case even if the thalassaemic chromosome is transferred into a mouse cell. These effects can be explained by the LCR opening up the chromatin so as to allow preferential DNA replication; however, it is also possible that transcription factors bound to the LCR might interact more directly with the replication machinery.

Not only does the transcription apparatus have a profound effect upon DNA replication, but the converse is also very much the case. High-resolution laser scanning confocal microscopy suggests that transcription stops at genomic sites which are undergoing replication. This conclusion is consistent with biochemical analyses carried out *in vitro*, which found that the passage of a replication fork will displace transcription factors and nucleosomes from DNA. This was first demonstrated using initiation complexes assembled on *Xenopus* 5S rRNA genes: such complexes were found to provide no obstacle to the replication apparatus; furthermore, footprinting revealed that none of the factors remained associated with the replicated template. These results have profound implications for gene regulation, because they suggest that the transcriptional phenotype of a cell is liable to change each time that it duplicates its DNA. Whether or not a particular gene is transcribed following replication may be decided by what assembles first, an active initiation complex or a repressive chromatin array; once established, either may be stable and could remain until disrupted by the next round of DNA replication. Competition between transcription factors and nucleosomes has been observed on replicating 5S rRNA genes in *Xenopus* egg extracts: rapid and efficient chromatin formation could silence the replicated genes, but small reductions in the efficiency of nucleosome assembly were sufficient to allow the expression of newly duplicated DNA. Thus, during each S phase there may be intense competition between histones and transcription factors for binding to a promoter. This provides a window of opportunity for either re-establishing or altering the set of genes that are actively expressed.

Under most circumstances, the pattern of gene expression is maintained between a parent cell and its progeny. For example, the daughters of a proliferating hepatocyte are likely to transcribe the same set of genes as their parent, unless there is a change in extracellular signals. Furthermore, chromatin structure can be stably maintained between generations. These facts present a quandary, given that DNA replication is believed to erase both nucleosomes and transcription factors from the template. A possible solution comes from the observation that the pattern of DNA methylation is preserved through replication, because the maintenance methyltransferase acts on the nascent DNA within the replication fork. As was described in Chapter 8, methylated CpGs are bound by the factor MeCP2 and this can recruit histone deacetylases which direct the assembly of inaccessible chromatin structures. Transcriptionally silent genomic regions which were heavily methylated prior to replication are therefore likely to be incorporated in the progeny cells into similar closed structures containing deacetylated nucleosomes. Conversely, genes that are active and undermethylated in the parent are likely to be found in under-acetylated open chromatin following replication. It is therefore plausible that the inherited methylation pattern plays a major role in maintaining the patterns of gene expression in successive generations of cells.

Transcription and DNA repair

Genes are subjected continuously to the harmful effects of environmental agents. One of the most important sources of genetic damage is the ultraviolet component of sunlight, which can cause adjacent pyrimidines to dimerize (Fig. 13.2). The resultant dimers obstruct transcription and can give rise to permanent mutations following replication. In some cases this can kill cells, whereas in others it can result in carcinogenesis or inborn errors. It is there-fore of considerable importance that damage to genes should be repaired efficiently.

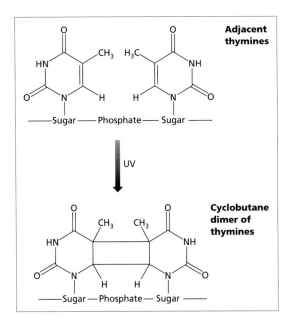

Fig. 13.2 Ultraviolet light can damage DNA by inducing the formation of covalent links between adjacent pyrimidines. The formation of a cyclobutane dimer of thymines is illustrated.

Chapter 11 described how the 'guardian of the genome' p53 helps protect higher organisms against mutation by blocking DNA replication and promoting its repair. In many cases this key transcription factor will trigger apoptosis in order to rid an organism of cells containing badly damaged DNA. Individuals who inherit inactive forms of p53 display a strong predisposition to cancer, a condition referred to as Li–Fraumeni syndrome. Tumour susceptibility can also result from defects in the machinery responsible for DNA repair. An example of this is provided by **x**eroderma **p**igmentosum (XP), a rare disease which results from mutations in the **n**ucleotide **e**xcision **r**epair (NER) apparatus. NER is the process whereby damaged DNA is removed as part of an oligonucleotide fragment, after which the intact strand serves as a template for the synthesis of replacement DNA. XP patients display extreme sensitivity to sunlight, pigmentation abnormalities and a more than 2000-fold elevated frequency of skin cancer, which is often accompanied by progressive neurological degeneration. This suggests that NER is a highly effective mechanism for protection against cancer. XP cells display hypermutability on exposure to UV radiation, which can account for the high frequency of sunlight-induced skin tumours. Defective NER is also associated with two other rare disorders that are not cancer prone, namely **t**richo**t**hio**d**ystrophy (TTD) and Cockayne syndrome. The defining feature of TTD is sulphur-deficient brittle hair, but this is associated with reduced size, mental retardation, ichthyosis (scaly skin) and in many but not all cases, sun sensitivity. Although TTD is a completely different clinical syndrome from XP, with few shared features, cells from XP and sun-sensitive TTD patients have indistinguishable defects in NER. Most TTD individuals have defects in the *XPD* gene, which encodes one of the helicase subunits of the basal pol II initiation factor TFIIH. Mutations in the gene encoding the other TFIIH helicase, which is called *XPB*, can also result in TTD and XP. TFIIH is the first basal initiation factor that has been implicated in human disease.

It is envisaged that the XPB and XPD helicases serve to open up the DNA duplex at the site of damage. They are thought to do this as part of TFIIH, rather than functioning independently in an unrelated repair complex. Thus, microinjection of purified TFIIH will allow *XPB* or *XPD* mutant cells to perform NER. Indeed, other TFIIH subunits besides the helicases have also been shown to be required for NER. However, NER can proceed unimpaired in the absence of the kinase components of TFIIH that phosphorylate the CTD of pol II. TFIIH is recruited to the repair sites by protein–protein interactions with XPA, a protein that recognizes damaged DNA using a zinc finger domain. The XPB and XPD subunits of TFIIH, which have opposite helicase polarities to each other, are thought to unwind the duplex so as to allow access of two structure-specific nucleases. These cleave the DNA on either side of the damage, leading to the excision of an oligonucleotide of about 29 bases. Repair synthesis by DNA polymerase ε fills in the resultant gap and the new DNA is joined to the old by DNA ligase (Fig. 13.3). Although this has been portrayed as a pathway

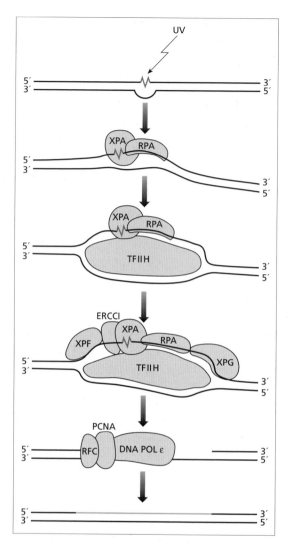

Fig. 13.3 Model of nucleotide excision repair (NER). The NER machinery involves the products of ~30 genes. It is responsible for removing oligonucleotides containing damaged DNA, allowing replacement of the missing sequence using the intact strand as template. The process has very broad specificity and will operate on DNA that has suffered a wide variety of chemical alterations. DNA damage is recognized by the zinc finger protein XPA and its affinity is increased through association with the single strand-binding protein RPA. XPA recruits TFIIH, which unwinds the damaged region by means of its helicase subunits XPB and XPD. XPA also recruits a heterodimeric complex composed of XPF and ERCC1. This has nuclease activity, as does XPG, which is recruited by RPA. The XPF/ERCC1 and XPG nucleases can cleave the damaged strand of DNA at the points where the unwinding begins and ends. This allows excision of the damage-containing oligonucleotide and release of the NER machinery. DNA polymerase ε, along with the replication accessory factors RFC and PCNA, can then synthesize a new oligonucleotide using the intact strand as template. This can be joined to the old DNA by DNA ligase I.

of stepwise recruitment, there is evidence that the proteins involved in the process of incision can aggregate into a large complex called the repairosome, which may be recruited in a single step as a preassembled entity. The concept of the repairosome is therefore directly analogous to the pol II holoenzyme described in Chapter 4, where many of the basal initiation factors are thought to preassociate before binding DNA.

Mutations in the *XPD* gene can result in patients suffering from either XP or TTD. Although there are no examples of an XP case having the same mutation as a TTD case, it has not been possible to distinguish an 'XP domain' and a separate 'TTD domain' within the XPD helicase. It is imagined that mutations which affect only the NER function of XPD will result in the XP phenotype, as the clinical symptoms of XP can be largely accounted for by a deficiency in DNA repair. However, the defining features of TTD, such as brittle hair and ichthyosis, cannot be explained readily by a lack of NER. TTD is therefore imagined to result from a subtle defect in the transcription function of TFIIH, which might affect selectively the synthesis of certain critical products, perhaps because they are especially dependent upon maximal rates of transcription. Examples might include the sulphur-rich proteins that are deficient in the hair of TTD sufferers. Mutations in XPD that affect transcription in this way may or may not also compromise the NER function, because TTD can occur with or without NER defects.

After pol II transcription has been inhibited by UV irradiation, it recovers faster than can be accounted for by the overall rate of damage removal that is observed for the genome as a whole. This is due to the preferential repair of genes which are transcribed actively by pol II. For example, in Chinese hamster ovary (CHO) cells the highly expressed *DHFR* gene is repaired five-fold more rapidly than the average for genomic DNA. Furthermore, the preferential repair is restricted to the template (transcribed) strand. Four hours after CHO cells have been irradiated, ~80% of cyclobutane pyrimidine dimers had been removed from the template strand of the *DHFR* gene, whereas most remained in the non-transcribed strand. These observations provided the first evidence that transcription might be coupled to some types of DNA repair and preceded the discovery that TFIIH is involved in both these processes. The targetting of repair in this way may be important because pyrimidine dimers in the template DNA strand serve as potent transcriptional inhibitors. Furthermore, the arrested elongation complex may remain stably bound at the site of DNA damage.

The dual role of TFIIH appeared to provide a mechanistic explanation of the coupling. If TFIIH remains associated with pol II during transcript elongation, it will be in immediate proximity to sites of DNA damage when transcription is arrested; it might then serve as a nucleation site for assembly of a complete NER complex. Although this model is appealing, several studies have found that TFIIH dissociates from pol II soon after transcription initiation. An alternative hypothesis involves TFIIH returning to pol II after it has stalled at a damage site, possibly because resumption of elongation might involve steps or specific conformations that occur during initiation. If TFIIH has affinity for arrested elongation complexes, it may serve to recruit the other components of the NER machinery. However, there is no direct evidence for this model and the biological significance has yet to be determined for the fact that TFIIH is involved in both transcription and repair. It is notable that the genes encoding large rRNA are not repaired preferentially, despite being transcribed at very high rates by pol I; TFIIH, of course, is not involved in the pol I system. Transcription-coupled repair is found in *E. coli*, where a coupling factor recognizes RNA polymerase molecules that have stalled at damage sites and then recruits the NER machinery, allowing the accelerated repair of the transcribed DNA strand.

The preferential repair of highly transcribed regions is defective in a condition referred to as Cockayne's syndrome (CS). The symptoms of CS include dwarfism, mental retardation, adipose deficiency, retinal atrophy, cataracts and acute sensitivity to sunlight. Unlike cells from XP or TTD patients, CS cells appear to be fully capable of mending DNA by the NER pathway; however, active genes are repaired at the same rate as silent genomic regions. It seems probable that the products of the two known CS genes are involved in coupling transcription to repair. This possibility is supported by the fact that one of these genes, *CSB*, encodes a protein with homology to the transcription repair coupling factor of *Escherichia coli*.

Somatic hypermutation and V(D)J recombination

In contrast to the above, genes encoding the variable (V) regions of immunoglobulins can be targetted for mutation in a process that appears to be linked to transcription. Somatic V-region hypermutation results in amino acid substitutions that increase considerably the diversity of antigen-binding sites within immunoglobulins. This process is restricted to a brief period in B-cell development, when they are in the germinal centre, and it generally requires activation by T cells. It results from non-templated single base changes in rearranged V-region genes. The frequency of these mutations is 10^{-3}–10^{-4}/bp/ generation, which is at least a million-fold greater than the rate of mutation in most mammalian genes and cells. Hypermutation begins just downstream of the promoter and ends after a further 1–2 kb. Heterologous DNA can serve as a target for hypermutation if it is substituted for the V(D)J sequences; this implies that the process is directed by *cis*-acting elements located outside the mutable region. Further evidence in support of this came from a study in which transgenic mice were created with targeted insertions in the H-chain locus; a transgene with a normal immunoglobulin promoter accumulated 15 times as many mutations as a promoterless transgene. A pol I promoter could also induce hypermutation in the transgene. It is currently unclear whether transcription itself is required for hypermutation, or whether gene accessibility or particular transcription factors are critical for the phenomenon. It has been suggested that the transcription apparatus may recruit a putative mutator protein that carries out error-prone repair at sites where the RNA polymerase pauses. At present, however, the subject remains highly speculative.

Transcription is also linked to the recombination process which joins the V, D and J segments of T-cell receptor and immunoglobulin genes together in different combinations to generate a large repertoire of receptors and antibodies. Thus, gene segments are transcribed prior to or coincident with their activation for rearrangement. Disrupting the enhancers that stimulate this transcription can often impair V(D)J recombination. Chromatin immunoprecipitation assays have shown that the regions undergoing recombination are associated with hyperacetylated histones. These observations are consistent with a model in which enhancer-bound transcription factors recruit histone acetyltransferases, which open specific chromatin regions and thereby allow access of the recombinase machinery.

Further reading

Reviews

Bentley, D. (1999) Coupling RNA polymerase II transcription with premRNA processing. *Curr. Opin. Cell Biol.* **11**: 347–351.

DePamphilis, M. L. (1993) How transcription factors regulate origins of DNA replication in eukaryotic cells. *Trends Cell Biol.* **3**: 161–167.

Drapkin, R., Sancar, A. & Reinberg, D. (1994) Where transcription meets repair. *Cell* **77**: 9–12.

Friedberg, E. C. (1996) Relationships between DNA repair and transcription. *Annu. Rev. Biochem.* **65**: 15–42.

Lehmann, A. R. (1995) Nucleotide excision repair and the link with transcription. *Trends Biochem. Sci.* **20**: 402–405.

Minvielle-Sebastia, L. & Keller, W. (1999) mRNA polyadenylation and its coupling to other RNA processing reactions and to transcription. *Curr. Opin. Cell Biol.* **11**: 352–357.

Neugebauer, K. M. & Roth, M. B. (1997) Transcription units as RNA processing units. *Genes Dev.* **11**: 3279–3285.

Steinmetz, E. J. (1997) Pre-mRNA processing and the CTD of RNA polymerase II: the tail that wags the dog? *Cell* **89**: 491–494.

Svejstrup, J. Q., Vichi, P. & Egly, J.-M. (1996) The multiple roles of transcription/repair factor TFIIH. *Trends Biochem. Sci.* **21**: 346–350.

Wiesendanger, M., Scharff, M. D. & Edelmann, W. (1998) Somatic hypermutation, transcription, and DNA mismatch repair. *Cell* **94**: 415–418.

Selected papers

Transcription and RNA processing

Birse, C. E., Minvielle-Sebastia, L., Lee, B. A., Keller, W. & Proudfoot, N. J. (1998) Coupling termination of transcription to messenger RNA maturation in yeast. *Science* **280**: 289–301.

Cho, E. J., Takagi, T., Moore, C. R. & Buratowski, S. (1997) mRNA capping enzyme is recruited to the transcription complex by phosphorylation of the RNA polymerase II carboxy-terminal domain. *Genes Dev.* **11**: 3319–3326.

Dantonel, J.-C., Murthy, K. G. K., Manley, J. L. & Tora, L. (1997) Transcription factor TFIID recruits CPSF for formation of 3′ end of mRNA. *Nature* **389**: 399–402.

Hirose, Y. & Manley, J. L. (1998) RNA polymerase II is an essential mRNA polyadenylation factor. *Nature* **395**: 93–96.

McCracken, S., Fong, N., Rosonina, E. *et al.* (1997) 5′-capping enzymes are targeted to premRNA by binding to the phosphorylated carboxy-terminal domain of RNA polymerase II. *Genes Dev.* **11**: 3306–3318.

McCracken, S., Fong, N., Yankulov, K. *et al.* (1997) The C-terminal domain of RNA polymerase II couples mRNA processing to transcription. *Nature* **385**: 357–361.

Transcription and DNA replication

Aladjem, M. I. *et al.* (1995) Participation of the human β-globin locus control region in initiation of DNA replication. *Science* **270**: 815–818.

Almouzni, G., Mechali, M. & Wolffe, A. P. (1990) Competition between transcription complex assembly and chromatin assembly on replicating DNA. *EMBO J.* **9**: 573–582.

Lee, D., Sohn, H., Kalpana, G. V. & Choe, J. (1999) Interaction of E1 and hSNF5 proteins stimulates replication of human papillomavirus DNA. *Nature* **399**: 487–491.

Wei, X., Samarabandu, J., Devdhar, R. S. *et al.* (1998) Segregation of transcription and replication sites into higher order domains. *Science* **281**: 1502–1505.

Wolffe, A. P. & Brown, D. D. (1986) DNA replication *in vitro* erases a *Xenopus* 5S RNA gene transcription complex. *Cell* **47**: 217–227.

TFIIH and nucleotide excision repair

Drapkin, R., Reardon, J. T., Ansari, A. *et al.* (1994) Dual role of TFIIH in DNA excision repair and in transcription by RNA polymerase II. *Nature* **368**: 769–772.

Schaeffer, L., Roy, R., Humbert, S. *et al.* (1993) DNA repair helicase: a component of BTF-2 (TFIIH) basic transcription factor. *Science* **260**: 58–63.

Svejstrup, J. Q., Wang, Z., Feaver, W. J. *et al.* (1995) Different forms of TFIIH for transcription and DNA repair: holo-TFIIH and a nucleotide excision repairosome. *Cell* **80**: 21–28.

Somatic hypermutation and V(D)J recombination

Fukita, Y., Jacobs, H. & Rajewsky, K. (1998) Somatic hypermutation in the heavy chain locus correlates with transcription. *Immunity* **9**: 105–114.

McMurray, M. T. & Krangel, M. S. (2000) A role for histone acetylation in the developmental regulation of V(D)J recombination. *Science* **287**: 495–498.

Chapter 14: Transcription Factors and Development

Development involves an orderly series of changes in the transcription patterns of genes. This is controlled by variations in the levels and activities of key transcription factors. Probably the best studied example is in *Drosophila* embryogenesis, where a hierarchical cascade of transcription factors is responsible for organizing the development of the fly. This chapter will describe the role of transcription factors in determining pattern formation along the anterior–posterior axis of the embryo in *Drosophila melanogaster* (Table 14.1). Many aspects of this system provide paradigms for other organisms, including mammals. It will also serve to illustrate in an important biological context some of the principles that have been met in previous chapters.

Bicoid

The organization of anterior–posterior polarity is initiated by a transcription factor called Bicoid, which binds DNA by means of a homeodomain. Maternal cells inject Bicoid-encoding mRNA into the anterior of an egg. This RNA remains near the site of entry, due to its interaction with microtubules, but when it is

Table 14.1 Some of the transcription factors involved in establishing the anterior–posterior axis of a *Drosophila* embryo.

Transcription factor	DNA-binding domain	Category
Bicoid	Homeodomain	Anterior morphogen
Caudal	Homeodomain	Posterior morphogen
Hunchback	C_2–H_2 Zinc fingers	Gap gene product
Kruppel	C_2–H_2 Zinc fingers	Gap gene product
Knirps	C_2–C_2 Zinc fingers	Gap gene product
Giant	bZip	Gap gene product
Hairy	bHLH	Pair-rule gene product
Even-skipped	Homeodomain	Pair-rule gene product
Fushi tarazu	Homeodomain	Pair-rule gene product
Paired	Homeodomain	Pair-rule gene product
Engrailed	Homeodomain	Segment polarity gene product
Gooseberry	Homeodomain	Segment polarity gene product
Armadillo	None	Segment polarity gene product
Antennapaedia	Homeodomain	Homeotic gene product
Ultrabithorax	Homeodomain	Homeotic gene product
Distal-less	Homeodomain	Synthesis repressed by Ultrabithorax
Homothorax	Homeodomain	Synthesis repressed by Antp
Extradenticle	Homeodomain	Synthesis repressed by Antp

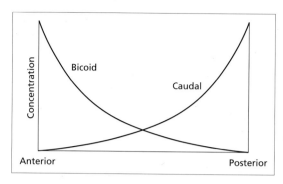

Fig. 14.1 Bicoid and Caudal proteins are distributed in opposing concentration gradients in the preblastoderm embryo. The highest concentration of Bicoid is at the anterior and the lowest at the posterior; this gradient forms because the *bicoid* mRNA is localized at the posterior, but its protein product can diffuse away. The highest concentration of Caudal is at the posterior and the lowest is at the anterior; this gradient forms because Bicoid binds to the evenly distributed *caudal* mRNA and suppresses its translation.

translated the Bicoid protein diffuses away. This gives rise to a concentration gradient that spans two to three orders of magnitude, with a high Bicoid concentration at the anterior and decreasing levels towards the posterior (Fig. 14.1). The gradient provides a morphogenetic signal, which allows nuclei to sense their position along the anterior–posterior axis. This mechanism is particularly effective at the beginning of fly development because the early embryo develops as a syncytium, where the nuclei divide without being separated by cell membranes. This feature allows ready access of a diffusing transcription factor.

Embryos derived from mothers that lack the *bicoid* gene have no real head or thorax, although their posterior is relatively normal. The defect can be rescued by injection of cytoplasm from a *bicoid*$^+$ egg, especially if that cytoplasm was taken from the anterior. These observations reveal clearly that Bicoid is critical for development of the head and thorax. The key importance of its concentration is illustrated by the effect of manipulating the number of copies of *bicoid* genes carried by the mother, resulting in steeper or shallower gradients in the progeny. Loss of one copy produces an embryo with a much smaller head, whereas extra copies can result in an enlarged head region that occupies much of the embryo. The dose of Bicoid received by a nucleus dictates whether it initiates a pattern of gene expression for head, thorax or abdomen development.

One of the key genes that is activated by Bicoid is *hunchback*. Indeed, *hunchback*$^-$ embryos resemble *bicoid*$^-$ embryos, because in both cases the anterior region develops abnormally. Bicoid makes use of its homeodomain to bind cooperatively to multiple sites in the *hunchback* promoter and thereby stimulate transcription of this gene. As a consequence, *hunchback* is expressed in a broad band in the anterior half of the early embryo, where Bicoid levels are highest. The zone of *hunchback* transcription can be extended towards the posterior by increasing the dose of Bicoid (Fig. 14.2). The posterior border of the initial *hunchback* expression domain is defined by a minimum concentration of Bicoid. Cooperative DNA binding results in concerted loading of Bicoid onto clustered recognition sites; this steepens the transition between the on and off states for *hunchback* activation and thereby sharpens the border of the expression domain. This provides a paradigm for how a single transcription

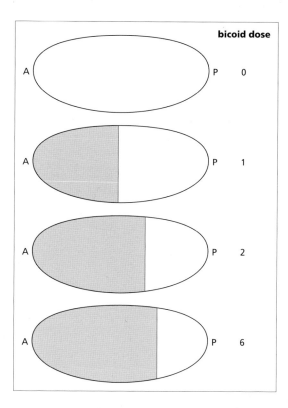

Fig. 14.2 Expression of the *hunchback* gene responds to the levels of Bicoid. The zone of *hunchback* expression within an embryo is indicated by shading. Embryos derived from *bicoid⁻* mothers do not activate the *hunchback* gene. In embryos from mothers with a single copy of the *bicoid* gene only half the normal dose of Bicoid is received and the zone of Hunchback activation extends less far than usual. By contrast, *hunchback* expression extends posteriorly when the Bicoid dose is abnormally elevated. A, anterior; P, posterior.

factor can convey positional information. Differential responses can be obtained to the same information if different genes have varying affinities for the transcription factor that forms the gradient. Thus, a promoter with low affinity binding sites for Bicoid will only be activated at the anterior of the embryo, where the Bicoid concentration is highest; by contrast, the low levels of Bicoid towards the posterior of the embryo might be sufficient to activate a promoter with very high affinity binding sites.

In addition to its role as a transcription factor, Bicoid performs an additional function as a specific regulator of mRNA translation. The posterior part of the early embryo is organized in response to a homeodomain-containing transcription factor called Caudal. Caudal protein forms a concentration gradient that is highest at the posterior and decreases towards the anterior, i.e. the converse to the Bicoid gradient (Fig. 14.1). Whereas the Bicoid protein gradient is achieved by diffusion from a localized source of *bicoid* mRNA, *caudal* mRNA is evenly distributed in the early embryo. However, Bicoid protein binds to the 3′ untranslated region of *caudal* mRNA and suppresses its translation. As a consequence, Caudal protein is absent from the anterior of the embryo, where Bicoid protein is most concentrated, and its level increases towards the posterior as the abundance of Bicoid falls. In effect, Bicoid serves to establish two opposing positional gradients, one emanating from each tip of the early *Drosophila* embryo.

Gap genes

As mentioned above, *hunchback* is one of the genes that is bound and regulated by Bicoid. *Hunchback* itself encodes a transcription factor, which uses zinc fingers to bind DNA. The Hunchback protein cooperates with Bicoid to regulate the *hunchback* promoter, providing another example of feedback control. As was described in Chapter 5, Hunchback and Bicoid can activate transcription synergistically via contacts they make with the $TAF_{II}110$ and $TAF_{II}60$ subunits of TFIID. Such cooperative interactions can give rise to sharp spatial limits for the expression domains of target genes.

Bicoid and Hunchback also activate a gene called *Kruppel*, which encodes another factor with a zinc finger DNA-binding domain. Despite being activated by Bicoid and Hunchback, *Kruppel* is not expressed in the anterior of the embryo. One reason for this is that *Kruppel* is repressed by high levels of Hunchback, although it is stimulated by lower levels. This results in *Kruppel* expression being restricted to the middle of the embryo; the anterior contains so much Hunchback that it represses *Kruppel*, whereas the posterior contains insufficient Hunchback to allow transcription of its target (Fig. 14.3). Another reason why *Kruppel* expression is limited to a central zone is that it is inhibited by the action of repressor proteins present in the anterior and posterior.

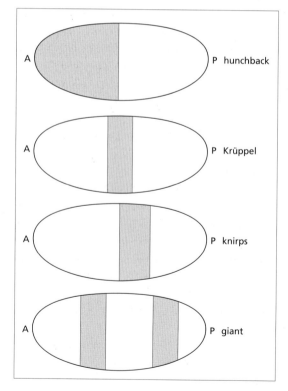

Fig. 14.3 Expression domains of the gap genes *hunchback, Kruppel, knirps* and *giant* in a blastoderm stage *Drosophila* embryo. Shaded regions indicate the domains where these genes are expressed, as determined by *in situ* hybridization to localize specific transcripts. A, anterior; P, posterior.

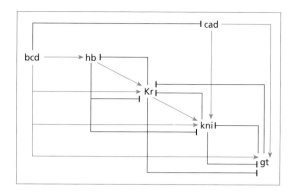

Fig. 14.4 Regulatory interactions between Bicoid (bcd), Caudal (cad) and the gap genes *hunchback (hb), Kruppel (Kr), knirps (kni)* and *giant (gt)*. An arrow indicates a stimulatory interaction, whereas a blocked line indicates repression.

Examples include the bZip protein Giant and the nuclear hormone receptor Knirps, which provide part of an inhibitor system that stabilizes and restricts the initially broad *Kruppel* expression domain that is established by Bicoid and Hunchback (Fig. 14.4). Giant and Knirps repress *Kruppel* transcription by antagonizing the stimulatory effect of Bicoid. A 730-bp region from the *Kruppel* enhancer contains six binding sites for Bicoid and allows activation by Bicoid in a concentration-dependent manner. One of these binding sites is overlapped by a binding site for Knirps, allowing Knirps to act competitively to exclude Bicoid and inhibit expression of *Kruppel*. Similarly, three recognition sites for Giant overlap with the Bicoid-binding sequences, providing further opportunities for competitive inhibition. By contrast, Hunchback binds to a separate cluster of sites within the *Kruppel* promoter in order to repress transcription.

Like Hunchback, Kruppel can act synergistically with Bicoid. Kruppel also resembles Hunchback in being able to function as either an activator or a repressor, in a concentration-dependent manner. When present at relatively low concentrations, Kruppel binds DNA as a monomer and stimulates transcription by interacting with TFIIB. *Knirps* is one gene that is activated by low levels of Kruppel; it is transcribed in a region that overlaps with the posterior of the Kruppel expression domain. However, at higher concentrations Kruppel can bind as a dimer to a single 11-bp DNA recognition site; under these circumstances it represses transcription, apparently through interaction with TFIIE and recruitment of a corepressor. The *knirps* expression domain is therefore prevented from spreading anteriorly by the increasing levels of Kruppel. This effect is reinforced through inhibition by Hunchback, which binds to at least three *knirps* sites. Similarly, Giant discourages the spread of *knirps* expression in both anterior and posterior directions. Unlike the overlapping binding sites found in the *Kruppel* enhancer, *knirps* is controlled by a modular array of separate DNA elements that allow activators and repressors to bind in parallel (Fig. 14.5).

Hunchback, Kruppel, knirps and *giant* are referred to as gap genes; if any one of them is deleted, its normal zone of expression fails to develop, giving the appearance of a gap in the mutant embryo. In each case, Bicoid is involved

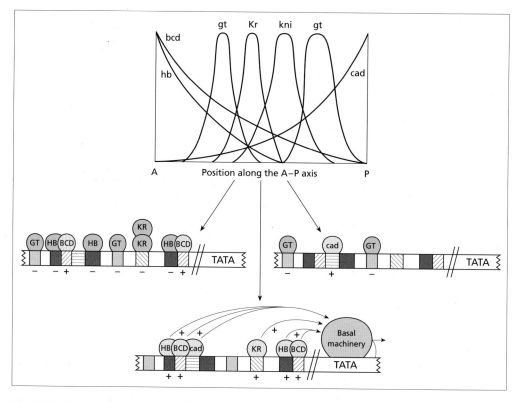

Fig. 14.5 Schematic illustration of how the occupancy of the regulatory regions of *knirps*
(kni) might vary according to position along the anterior–posterior body length, giving
rise to differential gene expression. The upper panel indicates where *bicoid (bcd)*,
caudal (cad), hunchback (hb), Kruppel (Kr), knirps (kni) and *giant (gt)* are expressed along the
anterior–posterior (A–P) axis of the developing embryo. The bottom panel presents a
hypothetical illustration of how the *knirps* regulatory region might be occupied at different
positions along the A–P axis. *Knirps* is transcribed in a broad band under the positive
influence of Bicoid (BCD), Caudal (CAD) and Kruppel (KR) monomers. Its expression
domain is prevented from spreading anteriorly by the repressive influence of Hunchback
(HB), Giant (GT) and Kruppel dimers. Despite the presence of Caudal, posterior expansion
of its expression domain is prevented by lack of its activators Bicoid and Kruppel and by the
presence of terminal repressors, one of which is Giant.

in activating transcription. A complex series of mutual interactions is then
responsible for defining and restricting the regions in which they are expressed (Fig. 14.4). These interactions establish bell-shaped concentration
gradients along the anterior–posterior axis that subdivide the embryo into
distinct spatial zones. There are additional gap genes which together give rise
to a dozen localized transcription factor gradients. The position of each cell
with respect to the anterior and posterior can therefore be defined by the
relative levels of Bicoid, Caudal and the various gap gene products that it
expresses. These combine to control the transcription of the next set of genes
in the regulatory hierarchy, which are called the pair-rule genes.

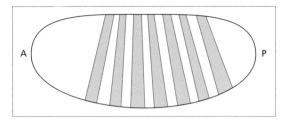

Fig. 14.6 Pair-rule genes are expressed in a pattern of seven evenly spaced stripes within the trunk region of the developing embryo. Shaded regions show zones of expression. A, anterior; P, posterior.

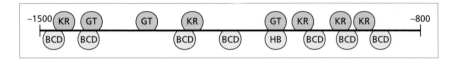

Fig. 14.7 Binding sites that have been mapped by footprinting within the enhancer that controls the second stripe of *eve* expression. Bicoid (BCD) and Hunchback (HB) serve to activate this stripe whereas Kruppel (KR) and Giant (GT) serve to limit its expansion. Many of the binding sites are found to overlap, allowing competition for occupancy.

Pair-rule genes

There are 10 pair-rule genes encoding transcription factors. Examples include *hairy*, which codes for a protein with a basic helix-loop-helix (bHLH) DNA-binding domain, and *paired*, *even-skipped* (*eve*) and *fushi tarazu* (*ftz*), each of which encode factors with homeodomains. The defining feature of the pair-rule genes is that they are each expressed in a pattern consisting of seven evenly spaced transverse stripes within the trunk region of the developing embryo (Fig. 14.6). The locations of the stripes vary between different pair-rule genes; for example, the stripes of *eve* are precisely out of register with those of *ftz*, so that trunk cells express either *eve* or *ftz*, but not both. The members of this group are essential for development, as shown by the fact that if a pair-rule gene is deleted, the resulting mutant embryo lacks those cells which would normally have expressed it (*fushi tarazu* means 'not enough stripes' in Japanese, and refers to the appearance of a *ftz⁻* mutant).

In the case of both *eve* and *hairy*, it has been shown that individual stripes of expression are governed by separate enhancer elements. For example, the enhancer responsible for the second *eve* stripe is located between 800 bp and 1500 bp upstream of the transcription start site. Transgenic flies containing a *lacZ* reporter gene linked to this 700-bp enhancer express the transgene in a single stripe that coincides with the second stripe of *eve*. Bicoid, Hunchback, Kruppel and Giant all bind to sites within this enhancer and control its activity (Fig. 14.7). Cooperative interactions between Bicoid and Hunchback are responsible for activation. Giant serves as a dominant repressor that prevents the stripe from spreading anteriorly. The posterior border of the stripe is set by the repressive action of Kruppel, as well as the decreasing concentration of Bicoid and Hunchback. In several cases, the binding sites for activators and repressors overlap with each other, resulting in competition for occupancy. These features are strongly reminiscent of the mechanisms involved in controlling the zones of gap gene expression. However, whereas *Kruppel* and *knirps* are transcribed in a single broad band, the pair-rule genes are expressed in seven separate stripes; *eve* and *hairy* utilize a sophisticated modular array of enhancer and promoter elements to generate this complex pattern, although the regulatory mechanisms at each element may be very similar.

Once the seven-stripe pattern is established, *eve* and *hairy* pass it on to some of the other pair-rule genes, such as *ftz* and *paired*. In principle this could be a much simpler process. For example, the action of Eve as a repressor to conteract some ubiquitous activator should be sufficient to establish the complimentary pattern of *ftz* stripes in only those cells which do not contain Eve. In fact, as so often, things are not that simple. A 600-bp regulatory region called the zebra element is found just upstream of the transcription start site of *ftz* and this is a target for several products of the primary pair-rule genes, which can act as either activators or repressors. The zebra element is also bound by the zinc finger protein Tramtrack, which was encountered in Chapter 3. Eve is, nevertheless, a key regulator of *ftz*. It inhibits transcription by at least two mechanisms: one involves recruiting a histone deacetylase, which is likely to reduce the accessibility of chromatin; the other mechanism involves direct contact with TATA-binding protein (TBP), which can prevent the recruitment of TFIID to promoters. Phosphorylation of Eve can block its ability to bind TBP and function as a repressor.

When they first arise, the stripes of *eve* and *ftz* expression appear quite fuzzy, which suggests that their borders are not rigorously defined. However, they soon become much sharper as they intensify. This reflects the action of autoregulatory feedback control loops.

Genes downstream of the pair-rule genes

Apart from regulating each other, the pair-rule transcription factors also control the expression of another set of targets called the segment polarity genes. These are transcribed in a pattern of 14 stripes and are required to distinguish between the anterior and the posterior of each segment of the developing fly. Mutations in these genes can change the polarity of every segment. Examples of this group include *engrailed* and *gooseberry*, both of which encode homeodomain proteins. *Engrailed* serves to define boundaries between the segments of the fly. Its zone of expression is specified so precisely by the pair-rule genes that the stripes are only a single cell wide when they first appear. The odd-numbered *engrailed* stripes are switched on by Paired whereas the even-numbered stripes are activated by Ftz. Other pair-rule factors help define the zones of expression by repressing *engrailed* transcription. The pattern is then maintained with the help of an autoregulatory feedback loop.

Not all of the segment polarity genes encode transcription factors. An important example is *wingless*, which encodes a secreted protein that mediates intercellular signalling, akin to the Wnt proteins in vertebrates. As was described in Chapter 10, Wnt triggers the accumulation of β-catenin by preventing its degradation; it then enters the nucleus, binds to TCF or LEF1 and coactivates transcription. A closely analogous system operates in *Drosophila*, where cells exposed to Wingless accumulate the β-catenin homologue Armadillo; the latter binds to *Drosophila* homologue of T-cell factor (dTCF) and activates a variety of target genes, including *engrailed* and *Ultrabithorax*. Indeed, mutations in *armadillo* can give a segment polarity phenotype that is similar to that seen in *wingless⁻* embryos. Conversely, overexpression of murine LEF1 in flies bypasses the need for Wingless signalling and produces a phenotype that mimicks a *wingless* gain-of-function mutation. The interaction between Armadillo and dTCF is regulated through the acetyltransferase activity of *Drosophila* CBP (dCBP). Although in most cases the HAT function of CBP (CREB binding protein) serves to stimulate transcription, as described by Chapters 5 and 8, in this instance dCBP has a repressive effect. Association of Armadillo with dTCF is inhibited when the latter is bound and acetylated by dCBP. This mechanism might serve to guard against inappropriate leaky expression of dTCF target genes; repression by dCBP may only be overcome when Wingless triggers an accumulation of Armadillo.

The dCBP cofactor is also involved in the function of another segment polarity gene product, called Cubitus interruptus (Ci). In the case of Ci, dCBP serves its more usual role as a coactivator and mutation of dCBP can prevent the expression of Ci target genes. The discovery that dCBP contributes to the normal function of Ci may have shed some light on two genetic disorders that occur in humans. One of these is Rubinstein–Taybi syndrome, which is caused by haploinsufficiency for the CBP gene and involves aberrant pattern development, including mental retardation and abnormal thumbs. This condition shows clear similarities with Grieg's cephalopolysyndactyly syndrome, which is caused by haploinsufficiency for a protein called Gli3. The latter belongs to the Gli family of mammalian transcription factors, which have zinc fingers and are homologous to Ci. Because Ci interacts with dCBP in *Drosophila*, it is quite plausible that Gli3 and CBP function together in humans. If they operate in the same pathway(s), this could explain the overlapping phenotypes observed when one copy of the Gli3 or CBP genes is lost.

As was the case for the gap and pair-rule genes, there is substantial cross-talk among the segment polarity genes. For example, Ci is involved in activating *wingless*, whilst Wingless stimulates expression of *engrailed*. Another target of the Wingless pathway is the homeotic gene *Ultrabithorax*, which is regulated by an enhancer with a binding site for dTCF. The homeotic genes are also subject to direct control by the pair-rule transcription factors. This group plays a key role in specifying the ultimate identity of groups of cells. For instance, mutations in homeotic genes can transform a thoracic segment into an abdominal one, or vice versa. More spectacular homeotic mutations can cause antennae to develop as legs or eyes to develop as wings. Inevitably, such dramatic changes in cellular phenotype involve the switching on or off of batteries of genes. It is therefore no surprise that the homeotic gene products are all transcription factors. In every case, they bind DNA by means of a homeodomain. Key examples are Antennapedia (Antp), which was described in Chapter 3, and Ultrabithorax. Highly conserved homologues, called Hox genes, are found in vertebrates, where they perform key functions in determining cellular identity. For instance, the homeodomains of *Drosophila* Antp and the human Hox-A7 protein differ in only one out of their 60 residues.

There is considerable evidence that the homeotic genes have contributed substantially to morphological diversity during insect evolution and it is probable that the same is true in higher organisms. In *Drosophila*, mutational disruption of *Ultrabithorax*, *Antp* and a third homeotic gene can cause all of the thoracic segments to develop the same, whereas they would normally each have a unique structure and appearance. This observation led to the idea that an ancestral fly might have had a series of identical thoracic segments, which subsequently diversified as homeotic genes evolved. One way in which the homeotic gene products promote the development of distinct segmental regions is by switching off the expression of genes required for other body parts. For example, a principle difference between the thoracic and abdominal segments is that the former carry legs whereas the latter do not. Limb formation requires a gene called *Distal-less*, which encodes a transcription factor with a homeodomain. In the developing abdomen, Ultrabithorax represses the expression of *Distal-less* by binding to one of its enhancer elements, thereby preventing legs from forming. Similarly, Antp has been shown to repress the expression of *extradenticle* and *homothorax*, homeobox-containing genes which drive the development of antennae. This can explain why loss-of-function *Antp* mutations can allow antennae to grow in place of legs and why ectopic expression of dominant *Antp* mutant alleles can transform the antennae into additional legs.

It is evident that many, but not all, of the genes regulated by homeotic proteins code for other transcription factors, such as *Distal-less*, *extradenticle* and *homothorax*. In some respects, this comes as no surprise, as the obvious way to determine cellular identity is by switching on and off particular sets of genes. It is nevertheless striking that we have now reached the fifth tier of transcription factors in the regulatory hierarchy that began with Bicoid (Fig. 14.8). Further tiers are likely, but it is not necessary to pursue them here, as the central role of transcription in development has now been clearly demonstrated. The pattern of a developing fly is dictated by a cascade of transcriptional decisions, which divide the embryo up into precise spatial domains and determine the phenotype of each cell within these domains. The fruitfly serves well to illustrate the extremely complex interplay of transcription factors that is required to orchestrate the embryogenesis of a relatively simple organism. In many

Fig. 14.8 A hierarchy of transcription factors is responsible for specifying the developmental fate of cells in a *Drosophila* embryo. The regulatory cascade is initiated by Bicoid, which controls the transcription of *caudal,* the gap genes and the pair-rule genes. The gap gene products then play a major role in determining the patterns of pair-rule gene expression. The pair-rule gene products, in turn, control the transcription of segment polarity and homeotic genes. The homeotic proteins continue the cascade by regulating another tier of target genes, some of which encode transcription factors such as *extradenticle* and *Distal-less*, which exert their effects by controlling the expression of further sets of genes. There is also substantial cross-talk between individual members of the gap, pair-rule, segment polarity and homeotic gene families.

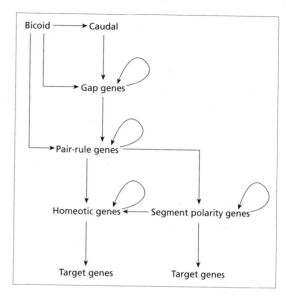

respects the development of a human follows similar principles to that of a fruitfly. The complexity is much greater, but the process is nevertheless based again upon the ability of key transcription factors to modulate the expression of genes.

Further reading

Reviews

Jackle, H. & Sauer, F. (1993) Transcriptional cascades in *Drosophila. Curr. Opin. Cell Biol.* **5**: 505–512.

Mannervik, M., Nibu, Y., Zhang, H. & Levine, M. (1999) Transcriptional coregulators in development. *Science* **284**: 606–609.

Morata, G. & Sanchez-Herrero, E. (1998) Pulling the fly's leg. *Nature* **392**: 657–658.

Rivera-Pomar, R. & Jackle, H. (1996) From gradients to stripes in *Drosophila* embryogenesis: filling in the gaps. *Trends Genet.* **12**: 478–483.

St Johnston, D. & Nusslein-Volhard, C. (1992) The origin of pattern and polarity in the *Drosophila* embryo. *Cell* **68**: 201–209.

Selected papers

Bicoid

Burz, D. S., Rivera-Pomar, R., Jackle, H. & Hanes, S. D. (1998) Cooperative DNA-binding by Bicoid provides a mechanism for threshold-dependent gene activation in the *Drosophila* embryo. *EMBO J.* **17**: 5998–6009.

Driever, W. & Nusslein-Volhard, C. (1989) The *bicoid* protein is a positive regulator of *hunchback* transcription in the early *Drosophila* embryo. *Nature* **337**: 138–143.

Rivera-Pomar, R., Niessing, D., Schmidt-Ott, U., Gehring, W. J. & Jackle, H. (1996) RNA binding and translational suppression by bicoid. *Nature* **379**: 746–749.

Struhl, G., Struhl, K. & Macdonald, P. M. (1989) The gradient morphogen *bicoid* is a concentration-dependent transcriptional activator. *Cell* **57**: 1259–1273.

Regulation of gap genes

Hoch, M., Gerwin, N., Taubert, H. & Jackle, H. (1992) Competition for overlapping sites in the regulatory region of the *Drosophila* gene *Kruppel*. *Science* **256**: 94–97.
Sauer, F. & Jackle, H. (1991) Concentration-dependent transcriptional activation or repression by *Kruppel* from a single binding site. *Nature* **353**: 563–566.
Sauer, F. & Jackle, H. (1993) Dimerization and the control of transcription by *Kruppel*. *Nature* **364**: 454–457.
Sauer, F., Fondell, J. D., Ohkuma, Y., Roeder, R. G. & Jackle, H. (1995) Control of transcription by Kruppel through interactions with TFIIB and TFIIEβ. *Nature* **375**: 162–164.
Struhl, G., Johnston, P. & Lawrence, P. A. (1992) Control of *Drosophila* body pattern by the *hunchback* morphogen gradient. *Cell* **69**: 237–249.

Regulation of pair-rule genes

Goto, T., Macdonald, P. & Maniatis, T. (1989) Early and late periodic patterns of *even skipped* expression are controlled by distinct regulatory elements that respond to different spatial cues. *Cell* **57**: 413–422.
Hiromi, Y., Kuroiwa, A. & Gehring, W. (1985) Control elements of the *Drosophila* segmentation gene *fushi tarazu*. *Cell* **43**: 603–613.
Li, C. & Manley, J. L. (1999) Allosteric regulation of Even-skipped repression activity by phosphorylation. *Mol. Cell* **3**: 77–86.
Small, S., Blair, A. & Levine, M. (1992) Regulation of *even-skipped* stripe 2 in the *Drosophila* embryo. *EMBO J.* **11**: 4047–4057.
Stanojevic, D., Small, S. & Levine, M. (1991) Regulation of a segmentation stripe by overlapping activators and repressors in the *Drosophila* embryo. *Science* **254**: 1385–1387.

The Wingless/Armadillo pathway

Riese, J., Yu, X., Munnerlyn, A. *et al.* (1997) LEF-1, a nuclear factor coordinating signalling inputs from wingless and decapentaplegic. *Cell* **88**: 777–787.
Waltzer, L. & Bienz, M. (1998) *Drosophila* CBP represses the transcription factor TCF to antagonize Wingless signalling. *Nature* **395**: 521–525.
van de Wetering, M., Cavallo, R., Dooijes, D. *et al.* (1997) Armadillo coactivates transcription driven by the product of the *Drosophila* segment polarity gene dTCF.*Cell* **88**: 789–799.

Targets of homeotic genes

Casares, F. & Mann, R. S. (1998) Control of antennal vs. leg development in *Drosophila*. *Nature* **392**: 723–726.
Vachon, G., Cohen, B., Pfeifle, C., McGuffin, M. E., Botas, J. & Cohen, S. M. (1992) Homeotic genes of the bithorax complex repress limb development in the abdomen of the *Drosophila* embryo through the target gene *Distal-less*. *Cell* **71**: 437–450.

Index